PYTHAGORAS'
REVENGE

ARTURO SANGALLI

PYTHAGORAS'
REVENGE

A Mathematical Mystery

Princeton University Press

Princeton and Oxford

Copyright © 2009 by Princeton University Press

Published by Princeton University Press, 41 William
Street, Princeton, New Jersey 08540
In the United Kingdom: Princeton University Press,
6 Oxford Street, Woodstock, Oxfordshire OX20 1TW
All Rights Reserved

ISBN: 978-0-691-04955-7

Library of Congress Control Number: 2008944149

British Library Cataloging-in-Publication
Data is available

This book has been composed in Minion

Printed on acid-free paper ∞
press.princeton.edu
Printed in the United States of America
1 3 5 7 9 10 8 6 4 2

PART III
A SECT OF
NEO-PYTHAGOREANS

PART IV
PYTHAGORAS' MISSION

Contents

Preface

When I first proposed the idea for this book to Princeton University Press, I had a completely different kind of project in mind. What was to be a reflection on the triumph and tyranny of numbers in modern societies became instead a work of fiction, a metamorphosis that was only possible thanks to the enthusiastic and unwavering support of Vickie Kearn, my editor at PUP.

I believed that through a fictional story I could more effectively attain my goal of introducing to a large audience certain mathematical concepts and results, some of them rather challenging and with philosophical undertones, in an entertaining way.

All I knew when I started working on the story was that it had to involve the figure of Pythagoras and his doctrines. Pythagoras wore so many fascinating hats—philosopher and mathematician, of course, but also religious leader, political and musical theorist, demigod, and miracle worker—that he appeared to me as the perfect pivotal character for a tale that would be part fact, part fiction, combining past and present, ancient beliefs and modern science, 2,500-year-old mathematics and the most recent advances in the field.

Pythagoras' Revenge should appeal to all those who enjoy reading about mathematics and mathematicians, from high school students to PhDs. In addition, by presenting mathematical ideas weaved into a suspenseful plot, I also hope to reach those who usually shun mathematics, and to initiate them to the beauty and the power of the "queen and servant of science," in the words of Eric Temple Bell.

Acknowledgments

During the writing of this book, I received technical advice from specialists in various fields to whom I would like to express my appreciation. Justin Croft, of Justin Croft Antiquarian Books, in Faversham, England, was a source of valuable information on the world of manuscripts and the rare book trade. Ted Stanley, Special Collections Paper Conservator at Princeton University Library and an expert in the conservation of rare paper-based objects, initiated me to the properties of papyrus and suggested some ancient techniques for the storage and preservation of papyri scrolls. Jerôme Morissette, an art conservator at the *Centre de Conservation du Québec*, generously shared with me some of his expert knowledge of the effect of time and the immediate environment on the structure of various metals. Roberta Partridge, book conservator at the Canadian Conservation Institute, directed me to some useful articles on book and papyrus conservation.

A number of people—colleagues, friends, and family—read earlier versions of the manuscript, and I greatly benefited from their generous comments, tactful criticism, and helpful suggestions. In no particular order, I would like to thank Andrew Watson, Aubert Daigneault, Roland Omnès, Frank O'Shea, Denys Cloutier, Gilles Plante, the anonymous (to me) PUP reviewers, and my daughter Natacha Sangalli, who also created the illustrations for the book.

A special *merci!* to my wife Francine Godbout, who helped me research background material, suggested a key location for the plot, supported me, encouraged me, and endured me throughout all the years it took for the book to materialize.

I am also indebted to all those at Princeton University Press who were involved in the production of this book, most especially to Vickie Kearn, who charted the course for the project and with a steady hand at the helm steered it to its successful conclusion.

List of Main Characters

(Chapter in which They Are Introduced)

JULE DAVIDSON, a mathematician (1)

JOHANNA DAVIDSON, Jule's twin sister; a consultant in computer security (1)

LEONARD RICHTER, alias Mr. Smith, a literary critic interested in enigmas (1)

ELMER GALWAY, a professor of classical history at Oriel College in Oxford (2)

BRADLEY JOHNSTON, a history of philosophy professor at Oriel and Elmer Galway's former student (2)

IRENA MONTRYAN, an assistant curator at the Royal Ontario Museum of Science (2)

JOHN GALWAY, Elmer's elder brother; a gem dealer (4)

DAVID GREEN, a rare book dealer (4)

ALFONSO LOPEZ DE BURGOS, a Spanish businessman (5)

LYSIS OF TARENTUM, a disciple of Pythagoras (8)

NORTON THORP, an internationally famous mathematician (9)

THERESE THORP, Norton's aunt and surrogate mother (9)

MORRIS PRINGLEY, Therese's close friend; an attorney (9)

ANDREW (ANDY) STONE, a retired professor of computer science (10)

GREGORY TRENCH, a doctor and member of The Beacon, an esoteric sect (14)

ROCKY, an ex-con; an acquaintance of Trench's (15)

HOUDINI, a computer wizard who works occasionally for Trench (15)

LAURA HIRSCH, a professor of classical studies; she's writing a book about Pythagoras (16)

Prologue

Pythagoras of Samos, the first man to call himself a philosopher—literally, a lover of wisdom—was a dominant figure in the history of ancient Greece. A brilliant mathematician and a mystic thinker, spiritual teacher, and political theorist, he embodies the intellectualism that was later to pervade classical Greek thought. The greatest Greek minds, from Aristotle to Proclus, all agree that he is the one who raised mathematics to the rank of a science.

But other, more obscure aspects of his personality cast some doubts on the true nature of the personage, such as his belief in the transmigration of the soul from humans to animals, his claim to divine status, and his alleged recollection of his previous reincarnations.

Pythagoras was born on the island of Samos, in the Aegean Sea, around 570 BC. As a young man he traveled to Egypt, where he was taught by the priests of Amon, the human-headed god of Thebes whose home was the temple of Karnak. He is also said to have met the naked philosophers of India before going to Babylon to study and teach astronomy, mathematics, and astrology.

When he was about forty years old he left Samos to escape the rule of the tyrant Polycrates and went to *Magna Graecia*, or Greater Greece, the name given to a group of Greek cities along the eastern coast of southern Italy. There he settled in the city of Croton, where he founded an ascetic and secretive sect. The fraternity, as their followers called it, was both a religious community and a scientific school devoted to exploring the mysteries of number, "the source and the root of all things."

For the Pythagoreans, "number" was a living reality whose nature was to be discovered. Their study of number was divided into four branches: Arithmetic, number in itself; Geometry, number in space; Music or Harmonics, number in time; and Astronomy, number in space and time. They believed that only through numbers may we achieve comprehension of things that would otherwise remain unknown; and that it is not only in all aspects of nature that we may see the manifestation of number but also in the creations of art and music.

Music played a central role in the Pythagorean doctrine. Pythagoras' discovery that harmonious sounds correspond to simple numerical proportions led him to extend this connection to the order of the universe at large and state that "the whole heaven or visible universe is a musical scale or number." He soothed the passions of the soul and body with the music of the lyre, by playing certain rhythms and singing certain songs that he composed.

Pythagoras taught the immortality of the soul, and that after death the soul transmigrates into other animated bodies. For this reason, all animated beings should be considered as belonging to one great family. He also taught that after certain specified periods the same events occur again, for nothing is entirely new; and that man is a microcosm, reflecting all the elements that make up the universe. He was the first to apply the word *kosmos* (literally, ordered-world) to the universe. But *kosmos* also means "ornament," so that according to Pythagoras the world is adorned with order.

Most of what we know about Pythagoras is shrouded in mystery, a mixture of fact and legend transmitted to us largely through the writings of Greek historians centuries after his time, since the earliest and most reliable accounts have for the most part been lost. These various sources often differ and sometimes openly contradict each other, as on the cause of his death: some say he died in a fire in Croton, others report that he survived the fire and fled to Metapontum, where he died of old age; according to yet another version, he was murdered by an angry mob. On one point, though, all ancient and modern historians agree: Pythagoras left no writings.

But what if he had? What if he had left a manuscript so well hidden that it was never found? Then, a flurry of questions would rush to mind: What was the manuscript about? Why did he write it? And why did he take such extraordinary precautions to preserve it?

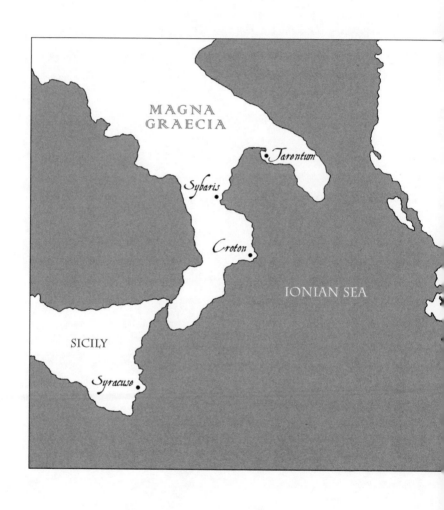

MAGNA
GRAECIA

Tarentum

Sybaris

Croton

IONIAN SEA

SICILY

Syracuse

THRACIA

MACEDON

PHRYGIA

Troy

ASIA
MINOR

Mount Olympus

AEGEAN SEA

GREECE

LYDIA

Delphi   Thebes

Athens

IONIA
Samos

Olympia

Sparta

PART I
A TIME CAPSULE?

Chapter 1

THE FIFTEEN PUZZLE

"Do you know the game called the Fifteen Puzzle?" asked the man, who had introduced himself simply as "Mr. Smith." Jule replied that he didn't think so. "It was invented around 1870 by Sam Loyd," the man went on, "one of America's greatest puzzlists, and at the time it became very popular, much as Rubik's cube did a century later."

Jule remembered his fascination with Rubik's cube as a teenager. The twenty-six brightly colored little cubes he had endlessly rotated searching for the elusive solution, a configuration with a single color on each of the cube's six faces. He also recalled wondering how many different arrangements were possible. Johanna, his twin sister, believed that the number of combinations was infinite, and also that once properly scrambled it was impossible to restore the cube to its original position. He knew that she was wrong on both counts, but was unable to prove his point back then. Not until many years later, long after he had forgotten all about the game and his fascination with it, did he come across the answer in one of the dozens of mathematical articles that Rubik's invention had spawned. "There are exactly 43,252,003,274,489,856,000 different configurations, not infinitely many," he had triumphantly announced to Johanna. But she would not quite concede defeat. "Well," she had said after a short moment of reflection, "such a big number is *almost* infinity."

The man reached into one of his pockets and produced a small wooden square with numbers on it. There was something vaguely familiar about the object but Jule could not quite tell what it was.

Loyd's Fifteen Puzzle starting position

The wooden square was in fact a frame containing smaller squares of identical size numbered from 1 to 15. These were arranged in rows of four, in sequential order starting at the upper left corner except for the last two squares, which were in reverse order, 15 followed by 14, leaving a vacant spot in the lower right corner. "The purpose of the game is to arrive at the sequential configuration 1 to 15 by sliding the squares up, down, right, or left into the empty spot one at a time," explained the man.

Jule thought he knew what would come next, but the man put the game back in his pocket. If he intended to challenge Jule with the puzzle, he did not appear ready to do it yet. Whatever the case, Jule was certain that sooner or later he would be put through some kind of test, given the circumstances leading to his presence at Mr. Smith's (or whatever his real name was) imposing two-story house in Highland Park, an affluent Chicago suburb of tree-lined curving streets and sprawling homes some twenty miles from the city center.

At age thirty-four, Jule Davidson had all but given up hope of ever becoming a famous mathematician. He was a rather short man with an athletic build and an easy manner. Despite his high forehead and thinning chestnut hair he considered himself good-looking, and you could grant him that, on account of his wide green eyes and well-proportioned nose. On the surface, he appeared to be reasonably happy with his teaching job at the Department of Mathematics of Indiana State University, in Terre Haute. But to his closest friends he would confess a growing discontent with the routine of an academic life that had become a little too comfortable and predictable, with its annual cycle of lectures, meetings, exams, and graduation ceremonies.

Jule secretly envied his twin sister. Johanna worked freelance as a consultant in computer security and was constantly traveling at short notice to London, Athens, or Bangkok to help some company keep two steps ahead of the next electronic intruder. And whenever a hacker gained access to sensitive data or the latest computer virus crippled one of her clients' systems, it was up to Johanna to fix the problem—which, unlike his course in group theory or non-Euclidean geometry, was never the same. He wished he too could put his mathematical talent and logical mind to some challenging use. In the daydreams of his youth, he had often pictured himself as the hero of some fantastic voyage or adventure of the kind imagined by Jules Verne and which the popular French writer had so vividly described in his novels, prophetically anticipating many of the twentieth century's technological marvels, from submarines to spaceships.

Actually, Davidson's first name was originally spelled "Jules," for he had been named after the famous author—"a literary genius and scientific visionary," in his mother's opinion. But he had later dropped the "s" because even if it is silent in French most people pronounced it, in effect changing his name.

One evening, while visiting canyousolveit.com, a Web site where mathematical puzzles were posted, a two-part question in probability had caught his eye, perhaps because it reminded him of a similar problem from his student days at the University of New Hampshire: *A group of twelve baseball players put their caps in a bag. After the caps are well shuffled, each player picks one at random. (1) Calculate the probability that none of the players will pick his own cap; (2) What is this probability if there are infinitely many players in the group?*

Not without some effort, Jule had found what he believed to be the correct answers.* For the first part of the question, he came up with 0.3679, or a 36.79 percent probability that none of the twelve players will pick his own cap. In order to answer the second part he had to find the limiting value of the probability as the number of players increases to infinity. Strange as it may seem, the probability remained practically the same regardless of the number of players. (For very few players, this is not the case; if there are only two or three players, for example, the probabilities are 0.5 and 1/3, respectively, as can be seen by a simple calculation.) Jule had calculated the probability that none of the infinitely many players would pick his own cap to be exactly $1/e$ or 0.367879441. . . .

The e appearing in the answer denotes a number of central importance in mathematics, a kind of universal constant. It is formally defined as the limit of the expression $(1 + 1/n)^n$ as n approaches infinity, but it is best known as the base of the natural logarithms. Its value is 2.7182 . . . (an infinite string of decimals follows). It crops up in a variety of contexts in pure mathematics and its applications to real-life situations, from the theory of complex numbers and differential equations to models of population growth, the arrangement of seeds in a sunflower, and probabilities involving baseball players exchanging their caps. Here's another example: If $1 is invested at an interest rate of 100 percent per year compounded every hour, the amount in the account after one year will be very close to e—that is, $2.72, rounded to the nearest cent—and it will be exactly e if the interest is compounded "continuously."

After Jule had clicked on the "Answer" button to check his solutions against those of whoever had posted the problem, a command had appeared prompting him to enter them in a box. He had done as instructed and the screen had responded with a message: *Now that you have passed the first test, would you like to go on? The prize is an opportunity to help solve a 2,500-year-old enigma.*

It had taken Jule three days to solve four other problems and answer a battery of short, nonmathematical questions such as: *What do the words "live," "record," and "lead" have in common?* and *Estimate the*

---

*For hints on the solution, see appendix 1.

number of words (on average) uttered by an American male during his lifetime, assuming he lives to age seventy-eight. After he had entered his answer to the last challenge—a fiendish mathematical puzzle—a fireworks display had filled the screen and a message had appeared: *Congratulations! You may be the person we're looking for. If you're still interested, please send resume to . . .*

Jule had e-mailed his curriculum vitae the next day. Even though he feared doing so could make him the target of endless spam, the possibility that it might lead to a change in his life was worth the risk.

One week later, in early January 1998, at two o'clock in the afternoon and after a tiresome six-hour drive from Terre Haute to the shores of Lake Michigan, he was calling at the Highland Park address sent to him with the invitation for an interview. A tall, slender man in his mid-fifties with penetrating brown eyes and silvery hair opened the door and greeted him with a smile. "Mister Davidson? Please come in." The man was impeccably dressed in a classic blue blazer, white shirt, and dark grey trousers, and was wearing a somewhat incongruous bow tie.

During the next hour or so, Jule was subjected to an extensive interrogation about his background, career, friends, and hobbies, and especially his motives for answering the enigmatic Internet message. And then, as if the interview had entered a new phase, the man had mentioned the Fifteen Puzzle and shown him the small square board on which the game was played.

"This puzzle has an interesting story," the man said, reaching for a book with a worn-out jacket that was lying on a low table. Jule observed the precise movements of the man's well-manicured hands, and noticed the silver ring with a facet-cut white stone on the middle finger of his left hand. Only weeks later would Jule discover the symbolic meaning of that ring.

"Would you like to hear the story?" asked the man, and without waiting for an answer began reading aloud in a theatrical tone. "In the late 1870s, the Fifteen Puzzle popped up in the United States; it spread quickly, and owing to the uncountable number of devoted players it had conquered, it became a plague. The same was observed on the other side of the ocean, in Europe. Here you could even see the passengers in horse trams with the game in their hands. In offices and shops bosses were horrified by their employees being completely absorbed by the game during office and class hours. In Paris the puzzle flourished in the open air, on the boulevards, and proliferated speedily from the capital all over the provinces. A French author of the day wrote, 'There was hardly one country cottage where this spider hadn't made its nest, lying in wait for a victim to flounder in its web.'"

There was a brief pause while the man glanced at Jule, as if to confirm that the story was having the effect he had anticipated. "In 1880," he went on, "the puzzle fever seemed to have reached its climax. The inventor of the game suggested to the editor of a New York newspaper to offer a $1,000 reward for the solution. The editor was a little reluctant, so the inventor offered to pay the reward with his own money. The inventor was Sam Loyd. He came to be widely known as an author of amusing problems and a multitude of puzzles. The $1,000 reward for the first correct solution remained unclaimed, although everybody was busy working on it. Funny stories were told of shopkeepers who forgot for this reason to open their shops, of respectful officials who stood throughout the night under a street lamp seeking a way to solve it. Nobody wanted to give up, as everyone was confident of imminent success. It was said that navigators allowed their ships to run aground, engine drivers took their trains past stations, and farmers neglected their ploughs."

At this point the man stopped reading and closed the book. "Do you know how the story ends?" he asked, fixing his eyes on Jule's. "I have no idea," Jule replied without hesitation. "Good," said the man, with more than a hint of relief in his voice. "Then we can proceed. Please follow me."

They entered a big room almost entirely filled with books. Most of them occupied the bookcases covering three of the four walls and many others were lying on a large table or in unsteady piles of various heights directly on the floor. The man led Jule to a computer terminal on their left, gestured for him to sit down, and then asked: "How about a little game of fifteen, Mr. Davidson?" Taking Jule's consent for granted, the man began to explain the rules. "The game board is a virtual one, as you might have already guessed." He pressed some keys and the puzzle's initial configuration appeared on the screen.

"The numbered squares can be moved around with the cursor—only permissible moves are allowed, of course. You will have exactly sixty minutes to solve the puzzle. Needless to say, my interest in retaining your services will greatly depend on the outcome of the game." He expected some reaction from Jule but there was none. "Do you have any questions?" he asked after a few moments.

Jule was already focusing on the task ahead and couldn't help remembering that for all his efforts he had never succeeded in solving Rubik's cube. But that was back in his youth, and now it was another story—and a different game. As his thoughts drifted toward Rubik's puzzle, he remembered reading that someone from Vietnam held the world record for solving the cube from a scrambled state in less than thirty seconds. He had not believed that possible. And yet, equally inconceivable feats were on record regarding the so-called calculating prodigies, individuals capable of performing complicated arithmetic operations in their head almost instantaneously but otherwise possessing an average intelligence. One of these, a certain Jacques Inaudi, born in Italy in 1867, was brought before the French Academy of Sciences at the age of twenty-five by the mathematician Gaston Darboux. They asked him questions such as: What day of the week was March 4, 1822? If the cube plus the square of a number equals 3,600, what is the number? Subtract 1,248,126,138,234,128,010 from 4,123, 547, 238,445,523,831. He had provided the correct answer to those and similar questions in less than thirty-five seconds.

Jule wondered how long it would have taken Inaudi to solve the Fifteen Puzzle. "What's the fastest that someone took to find the solution?" he wanted to ask. But the man was already gone and the screen was informing him that the game would start in 25 seconds, . . . 24, . . . 23, . . .

Chapter 2

THE IMPOSSIBLE MANUSCRIPT

"I'm afraid that would be impossible," replied Johnston in a dry tone. That was not the answer Irena wanted to hear. She had made the trip to Oxford expecting to see Dr. Elmer J. Galway, professor of Classical History at Oriel College and a specialist in pre-Socratic philosophy, but instead of meeting the famous scholar she was talking to Bradley Johnston, a junior member of the faculty who had been trying his best to fulfill her expectations, so far with little success.

Professor Galway had cancelled their appointment at the last moment and asked his colleague and former student to look after his overseas visitor. "A Miss Monyan, or Morian," he had told Johnston over the phone, adding that she worked for a museum in North America but giving no further explanations. It was early morning on a Thursday of November 1997 when Johnston received the call, and the appointment had been set for two o'clock that afternoon. He could not refuse Elmer such a favor, even if it meant missing cricket practice.

"Bradley Johnston," he introduced himself, "you must be Miss Morian." "Montryan," she corrected him, "Irena Montryan. Pleased to meet you. I have an appointment with Professor Galway." Johnston apologized on Elmer's behalf for the last-minute cancellation and led the way to the office, situated in one of the upper floors. He had not expected to see such a young woman, having perhaps unconsciously associated "museum" with "old." Her accent was definitely not British; that much he knew.

"Are you American?" he asked on the elevator. "Canadian," she corrected him for the second time. "But my father was American, from Minnesota. He emigrated to Canada after the war." "The *second* world war, I presume," Johnston said, venturing a smile. "Of course!" she exclaimed, returning the smile. "How old do you think I am?" They were both laughing as the elevator doors opened on the fourth floor.

They entered Johnston's office, a small room with a single tall window overlooking a central court known as the quadrangle. Oriel, founded in 1326, is one of Oxford's oldest colleges. Nothing survives of the original construction, a large house known as La Oriole that was demolished when the four buildings enclosing the quadrangle were built during the seventeenth century. Demand for more accommodation for undergraduates resulted in two more blocks being added a century later. The passage of time had had little effect on their austere exterior, but extensive interior renovations had taken place in order to increase office and classroom space and to accommodate some modern amenities. In the 1980s, a fourth floor and an elevator were added to one of the west wings where Johnston's office was located.

He offered her tea, but since she declined, he decided against making a cup only for himself. It was time to get down to business. "So what precisely did you want to see Professor Galway about, Miss Montryan?" asked Johnston.

"I work for the Royal Ontario Museum of Science in Toronto as assistant curator," she began. "As you may know, UNESCO has declared the year 2000 World Mathematical Year." He did not know but kept silent. "The ROMS is preparing an exhibition on the history of mathematics," she went on. "One of the main themes will be Greek mathematics, and in particular the Pythagoreans, given that mathematics as a science was born with the Pythagorean school. In fact, we are planning our marketing campaign around the figure of Pythagoras as an icon, the archetype of the philosopher-mathematician of ancient times."

"I see . . . , a kind of 'Pythagoras Superstar' that would appeal to the masses."

"That's precisely the idea. We are counting on Pythagoras to sell the exhibition to the general public. Who has not heard of the Pythagorean theorem?"

"Indeed. And since Elmer, I mean, Professor Galway is a leading authority on pre-Classical Greece . . . "

She finished the sentence for him: " . . . I came to seek some information from him. You're right again, Dr. Johnston."

"Please call me Brad. As a matter of fact, Irena . . . may I call you Irena?" She nodded. "As a matter of fact, I may be able to help you. I did my doctoral dissertation on Aristotle, whose treatises are considered the most reliable source on Pythagoreanism, in particular his *Metaphysics* and his monograph *On the Pythagoreans*, of which unfortunately only a few fragments have survived." He was beginning to warm up. The prospect of talking about his work to an attentive and attractive young lady had produced a stimulating effect on him and he no longer regretted the missed cricket practice.

But his excitement was to be short-lived. "Actually, Brad, I did not come to see Professor Galway to be lectured on the Pythagoreans and their mathematical theories—the museum has its own consultants for that. I hoped he could help me locate some artifacts, manuscripts perhaps, dating from Pythagoras' own time. Our exhibition will be centered on objects, and I thought it would be fantastic to be able to exhibit an original Greek mathematical manuscript, perhaps one written by Pythagoras' own hand. It would be the centerpiece of our exhibition."

"I'm afraid that would be impossible," was Johnston's dry reply. She noticed a shift in his mood. "Why's that?" she asked, raising her voice for the first time. "First of all," Johnston began to explain in a calm voice, "because written accounts were strictly forbidden. The Pythagoreans lived in a sort of brotherhood or sect devoted to the study of numbers, which they believed held the key to unraveling the mysteries of the universe. Their discoveries were considered sacred, and were passed on to other members of the group orally, under an oath of secrecy. Only when the few surviving Pythagoreans were expelled from southern Italy, in around 450 BC, did they preserve some of their knowledge in writing, according to certain sources. Pythagoras himself left no writings. All specialists in the field agree on that point." He then delivered the final blow to her hopes. "And even if the experts are wrong and Pythagoras did put down some of his teachings in writing, the probability that such a document, most likely a papyrus scroll, has survived for 2,500 years is virtually nil."

"How about the Dead Sea scrolls discovered in 1947?" She would not give up so easily. "They are from the second century BC, which makes them almost 2,200 years old." He thought for a moment before replying. "The exact age of those scrolls is still being debated," he said at last. "Carbon-14 dating put their origin at the year 68 of our era, which means they would be less than 2,000 years old. Besides, those scrolls were exceptionally well protected, wrapped in cloth and kept in pottery jars, with covers tied over the jar tops. On top of that, if you forgive my pun," he was beginning to enjoy being challenged, "we are talking here about a site in the Judean desert and therefore extremely dry and hot, the ideal conditions for preservation."

How does he know that the Pythagoreans didn't also take "exceptional precautions" to preserve their writings? she thought, but instead of putting the question to him she tried a different approach. "You say there is almost no chance that very ancient manuscripts have survived. Where did you learn all you know about Aristotle and his philosophy, then?" It was a simple question, but he appeared a little disconcerted by it. "Where did I learn it? Well . . . , from books, articles, lectures, scholarly reports, doctoral dissertations, discussions with other specialists and the like. The present body of knowledge on the classics is very impressive, you know." "I see," she said, not very convincingly. "And where did the authors of those learned articles and books get *their* knowledge? From other books?" He saw where she was leading him.

"What you really want to know is how it all started, don't you? What is the source, if there are no originals around any more?"

"Right." She was smiling, and enjoying having him on the defensive.

"It is a fact that none of the original works of Aristotle has survived," he said, "but there are copies, or rather copies of copies. Take Aristotle's *Physics*, for example." He got up, took a book from one of the shelves lining two walls of his office, and returned to his chair.

"This is an English translation by a highly respected philologist of the first two volumes of his *Physics*," he said, holding the book in one hand, "but by no means is it a translation of the original written by Aristotle himself—or dictated by him to one of his scribes.

"Before the author could begin the actual translation, he had to establish a text that was to the best of his judgment as close as possible to the original, using the scores of extant manuscripts of Aristotle's works, the oldest of which date from the ninth century—more than twelve hundred years after Aristotle's time. These 'sources' are all different and far from original. They have been copied from a variety of 'models' that are lost. And copyists make mistakes, sometimes unintentionally but other times with the best of intentions, to correct a presumed error in the model. Now, it is this 'established text'—a creation of the philologist out of 'copies of copies'—that finally gets translated." He stopped for a brief moment before asking: "Do you get the idea?"

She nodded and replied: "Yes, I see what you mean."

"The knowledge of the classics is therefore based on copies," he continued. "The oldest extant of these are written on papyrus, fragments of which, and occasionally entire rolls, survived, protected by the dry Egyptian soil." He was warming up again. "During the dynasty of the Ptolemies, when the Macedonian kings ruled in Egypt for three hundred years until 30 BC, it was customary to decorate human mummies with layers of linen cloth and papyrus fixed together with glue and coated with stucco. By removing the plaster, the layers could be separated and the documents used could be recovered in a more or less legible condition."

He pushed back his chair and crossed his legs before resuming. "In 1899, a team excavating at a Ptolemaic cemetery in Upper Egypt came across the *Athenaion Politeia*, Aristotle's treatise on the Constitution of Athens, until then known only by a few citations. It was one of the most sensational discoveries of those excavations. The four rolls making up the manuscript are palimpsests, papyri from which the original writing, a record book from a farm in the years 78 and 79, had been scraped off and written over with Aristotle's text. But for most of his works we have not been so lucky, and must rely on much more recent medieval copies that may contain many errors. Cicero speaks of 'books full of lies,' the work of unscrupulous copyists, or the result of using teams of low-paid, unskilled scribes who wrote from dictation."

"And where are those 'copies of copies'—the most ancient versions, I mean—kept?"

"Most have been bought by, or donated to, museums and universities. We have some right here in Oxford, at the Bodleian Library, and Aristotle's Constitution of Athens is at the British Museum. Others have found their way into private collections or to dealers in rare books and antiquities in London, Berlin, Vienna, or New York. There is a lucrative market for that kind of thing, and for every authentic document there are perhaps half a dozen forged ones."

Irena was beginning to appreciate how different the research methods in classical studies were from those in the natural sciences. She had a Master's degree in biology, and had done some research on the effect of pollutants on salmon reproduction before joining the "Save the Earth" movement and later the museum. Scientists look at facts and formulate theories, which their experiments will either confirm or refute. They have no practical use for old manuscripts, despite their sentimental or historical value. Classical scholars, on the other hand, do not start with facts but with words in ancient languages—which in many cases they only half understand—written on fragments of parchment or papyrus. Out of those scarce and often contradictory or unreliable sources, they have to piece together plausible interpretations and tentative hypotheses—hypotheses that cannot be tested by experiments.

She kept silent; her earlier enthusiasm had all but disappeared. "Are you telling me," she resumed wearily, "that I'll be wasting my time trying to find an original Pythagorean manuscript for my exhibition?" He felt a bit sorry for her and wished he could leave her on some optimistic note. "I can't . . . , I mean, as far as I know, yes. But I'm not an expert in that department. I wish Professor Galway were here. I'll talk to him as soon as he comes back and explain the situation. He'll most certainly get in touch with you. Do I have your card, by the way?" She handed him her business card over the cluttered desk and got up, preparing to leave.

"Would you join me for dinner?" he asked, getting to his feet. "There is an excellent Italian restaurant five minutes away, or we could go to a pub for a taste of the local cuisine."

"I'm sorry, I can't stay. I'm driving back to London right away; I have a dinner appointment there already."

"Oh, I understand," he said, trying not to appear too disappointed.

He accompanied her back to her rental car, a brand new red Opel. "Have a good trip, and remember to drive on the left side of the road," he said with a smile as they shook hands, and, almost apologetically, he added: "I wish I could have been more helpful." "Who knows?" she replied in a cheerful voice, "Maybe some mysterious private collector has been hoarding away some precious manuscripts no one has heard of yet." She too wanted to part on a positive note. "I'll try posting an ad on the Internet and see what happens."

"You never know," he said, not sounding very encouraging. He wished her good luck and waved to her as she drove off, the late afternoon sun reflecting on the rear window of the car.

Johnston was forced to admit that Irena had come to knock at the right door in her search for information on Greek manuscripts. Elmer was a leading scholar on the Archaic period of Greek civilization, going from the eighth to the sixth century BC, with a particular interest in the early Pythagoreans. He was also an accomplished linguist and philologist whose mastery of ancient languages had allowed him to study original sources and to author many translations from Phoenician and Greek that became standard references among historians. But Elmer was above all a field man, a man of action who had traveled the world. He was often away from Oxford, as a guest speaker at some international conference or visiting an archaeological site, when he wasn't responding to a call to give his expert advice on some ancient artifact or manuscript. If you needed to track down a document on ancient Greek history, he was your man.

Elmer's father, Sir Ernest, still in pretty good health at age ninety-three, was a famous archaeologist and explorer. He was only eighteen when he took part in the Antarctic expedition under the command of an even more famous Ernest—Sir Ernest Shackleton. It was to be Shackleton's last trip, for he died of a heart attack on board his ship, the *Quest*, in January 1922. Young Ernest's passion for adventure would later take a heavy toll on his family life. In 1931 he married Elizabeth Jennifer Williams, the only daughter of a barrister, who gave him two sons: John Arthur, in 1932, and Elmer James, born two years later.

Ernest was away most of the time digging at some remote site in Egypt, Greece, or South America, while Jenny looked after the children in their spacious Salisbury house. The boys did not get to see much of their father, except during the short intervals between his trips. Ernest would then take them to nearby Stonehenge for a visit and a lecture on the possible origin of the enigmatic monuments, or the whole family would spend a rare vacation together at the seaside in Bournemouth.

Not long after the war, in 1949, Ernest and Jenny divorced. It was a painless and long anticipated event. The boys were already away at boarding school when the Salisbury house was put up for sale. Besides its centuries-old massive furniture, it contained a vast collection of objects and artifacts that Ernest had brought back—or smuggled—from excavation sites around the world: sculptures, pottery, tools, weapons, religious vases, papyrus scrolls, coins, gold jewelry, and even several sarcophagi. Some of the objects were put in temporary storage while Ernest searched for new living quarters. Much to Elmer's regret, though, most of his father's treasures were donated or loaned to museums or sold at auction, but he later managed to recover a few items. Unlike his elder brother, he had always demonstrated a keen interest in those ancient artifacts, silent witnesses to the early chapters of humankind's unfolding saga.

Johnston had been invited to Elmer's place on Blackhall Road on several occasions, first as a student and later as a colleague. One evening after dinner, the professor had shown him and the other guests his coin collection, which he kept in velvet-lined trays inside a large cabinet in one corner of his study. Then Elmer had proudly called their attention to his favorite piece, dating from around the fourth century BC: a one-and-a-quarter-inch silver coin, depicting on one side the naked figure of a Greek god or hero, probably Heracles, holding a club in his left hand and a sort of chain in his right one, and sitting on the lion he had just conquered. The Greek words "KPOTΩNI ATAN" (coin from Croton) were written in big capital letters along the circular edge of the coin.

It was funny he should remember that particular event just now, he thought. Croton is the city in Italy where Pythagoras established his brotherhood, and Irena had come to inquire about Pythagoras.

He hoped Elmer would be able to help her. Johnston recalled his telephone call that morning. The professor appeared to be in a hurry and had vaguely mentioned "a business trip." A week later, Johnston would learn that Galway's trip also had something to do with Pythagoras.

Chapter 3
GAME OVER

The words "Game Over" appeared on the screen in big, flashing characters. The man who wished to be called Mr. Smith but whose real name was Leonard Richter had been following the game on a monitor from his private study. Not that there had been much to follow. Unlike most of the other candidates who had battled until the very end trying to solve the puzzle, Jule had stopped playing after only a quarter of an hour, so the screen had displayed what looked like a frozen image of the board for the remaining forty-five minutes.

Richter had first feared a system failure, but the screen clock had continued to operate normally. A more likely explanation was that Jule had used the time trying to figure out some principle or strategy that would lead to the solution, instead of playing the game by trial and error in the hope of hitting on the winning configuration by some lucky combination of moves. If such was the case, he had obviously run out of time. Unless . . .

When Richter entered the big room, Jule swiveled round in his chair to face him. He was smiling as he said: "Congratulations, Mr. Smith. That was a very neat trick you pulled on me. I wonder how many candidates for whatever you have to offer fell into the trap."

"Congratulations to you, Mr. Davidson." Richter also smiled, seemingly very pleased by what he heard, and completely ignoring Jule's sarcastic tone. "When did you find out that the puzzle had no solution?"

"Less than five minutes ago."

"And may I ask how you did it?"

"It happened while I was trying to understand the puzzle in mathematical terms. I discovered that it was impossible to reach certain configurations of the board by making only the permissible moves, that is, by sliding blocks into the vacant square. At that point it occurred to me that the desired final configuration, with the numbers in sequential order, might be among these. After all, your little story about the game never mentioned anyone who had actually solved the puzzle, even after the cash prize was offered."

"So you only *suspect* that the puzzle has no solution, but you don't know that for sure, do you?" Richter had expected a reasoned explanation, not just an educated—and lucky—guess.

"Not at all. I am certain that there is no solution, and I can prove it mathematically."

"I'd be glad to hear your proof, then."

"Are you a mathematician?"

Richter hesitated before replying. "I have degrees in philosophy and literature, but I've always been fascinated by enigmas, mathematical or otherwise, and have studied them for many years. Most of the books you see here deal with the subject, in one form or another. As for this particular puzzle, I am familiar with one mathematical proof of the impossibility of solving it, so I'm pretty sure I shall be able to understand yours, Mr. Davidson."

Jule reached for his notebook, which was lying on the table beside the computer monitor. He had been scribbling some notes but, with time running out, had not been able to check every link in the logical chain of his argument.

"The fundamental concept in play here is that of a permutation," he began. "Every configuration of the board, when read from left to right, top to bottom, becomes a permutation of the sixteen numbers 1 to 16, that is, a list of these numbers in a particular order (the empty square may be considered as the number '16'). For instance, the initial configuration becomes 1 2 3 4 5 6 7 8 9 10 11 12 13 15 14 16. The total number of such permutations is quite large, about 7 trillion."

"Or exactly 16 factorial," interrupted Richter, "that is, 16 times 15 times 14 times 13 and so forth, all the way down to 1. The reason for this being that there are sixteen choices for the first number in a permutation and, after it has been chosen, fifteen choices for the second, so there are 16×15 ways of choosing the first two numbers; $16 \times 15 \times 14$ ways of selecting the first three, and so on, until the 16th position can only be filled with the remaining number. This gives a total of $16 \times 15 \times 14 \times \ldots \times 2 \times 1$—'16 factorial,' in mathematical jargon—possible permutations of the sixteen numbers."

"That's right," confirmed Jule, wondering whether the man he knew only as Mr. Smith was trying to show off.

"Put in the simplest terms, a solution of the puzzle consists in transforming the initial permutation into the final one by a succession of moves of a particular kind: sliding a block (left, right, up, or down) into the adjacent empty square. Both the initial and the final permutations end with '16' (i.e., the empty square is in the lower right corner). Now, here is the first significant fact: To go from one permutation ending with '16' to another of the same type requires an even number of moves."

"And is that fact supposed to be obvious?" asked Richter.

"Not really, but I can easily convince you why it must be true. First notice that after each move the empty square itself 'moves' one square from its previous position on the board, either horizontally (left or right) or vertically (up or down), so that any series of moves causes the empty square to 'travel' around the board. Do you follow me?"

"Perfectly. Please go on." Richter appeared impatient to hear the rest of the proof. Jule resumed his explanation. "Suppose that you start from the initial configuration S and, after making a certain number of moves, you reach the final configuration F. Since at the end of its 'journey' around the board the blank square comes back to its original position (the lower right corner), it must have moved as many times up the board—m times, say—as it did down the board, so it must have moved vertically an even number ($2m$) of times; similarly, it must have moved horizontally also an even number of times. Hence, the total number of moves that took you from S to F, being the sum of two even numbers, must itself be an even number. We have thus established fact number 1: *Reaching the final configuration requires an even number of moves.*" At this point Jule paused and began checking his notes.

Richter's voice filled the momentary silence. "But that fact does not rule out the possibility of reaching the final configuration F, does it?"

Jule looked at him, slightly annoyed by the remark, and replied: "Of course it doesn't, not fact number 1 by itself. But wait until I prove to you fact number 2: *Reaching the final configuration requires an odd number of moves.* Then, by putting the two facts together, we must conclude that the Fifteen Puzzle has no solution: for if the puzzle could be solved in *n* moves, say, this number *n* would be both even and odd, a number that doesn't exist!"

"That would certainly clinch the argument," Richter commented enthusiastically, "I'm eager to hear your proof of fact number 2."

"And I shall be happy to oblige, but first I need a few minutes to review my notes, if you don't mind."

"Not at all. Take as much time as you need." And, getting on his feet, he added, "If you would excuse me, I have some business to attend to. It won't take long."

While Jule turned his attention to his notes, Richter left the room and headed back to his study. He closed the door behind him, sat down at his desk, and reached for a telephone half buried under a pile of papers. The telephone at the other end of the line rang several times before the answering machine was activated: "*We cannot take your call just now. Please leave a message.*" After a short beep, the machine began recording Richter's message: "Richter. I've got good news. I've found him. The team is now complete. Can you call me back this evening? Thank you." And he hung up.

"We need to go back to permutations," announced Jule to his audience of one. Richter was once again in the library room, sitting in a comfortable leather sofa, three-quarters of which was covered with books and sheets of paper.

"There are two sorts of permutations, even permutations and odd ones. Are you familiar with these concepts?"

"I'm afraid not," admitted Richter.

"Then we shall start from the beginning. Any given permutation can be 'unscrambled,' that is, restored to the natural order, 1 2 3 4 5, etc., by performing a certain number of 'exchanges.' Here's an example—we shall use six numbers instead of sixteen to make it simpler, but the principle is the same. You'll need some paper to write on." Richter picked up a blank sheet lying nearby and produced a fountain pen from one of his pockets.

"Please write '3 6 2 1 5 4.'" Richter did as requested, in small, blue numerals. "Now exchange 3 and 1 and write down the resulting permutation." The permutation '1 6 2 3 5 4' appeared on Richter's sheet, one numeral at a time, below the previous list. He dutifully continued to record each new permutation that resulted from the exchange dictated by Jule. "Now exchange 2 and 6 ('1 2 6 3 5 4'); 6 and 3 (1 2 3 6 5 4) and finally 6 and 4. You must now have arrived at 1 2 3 4 5 6, that is, the natural order."

"I have indeed," confirmed Richter, looking up from the sheet of paper.

"Notice that you needed four exchanges to unscramble the given permutation. Actually, there are many other sequences of exchanges that would achieve the unscrambling, but they all have something in common: the number of exchanges is always even—this is a well-known property of permutations that you can check out in any mathematics textbook on the subject. For this reason, we say that permutations such as 3 6 2 1 5 4 are *even*, while those requiring an odd number of exchanges to put them back in the natural order are said to be *odd*."

Jule paused, as if inviting Richter to ask questions, but his host kept silent.

"Now, regarding our puzzle, the initial configuration of the board is an odd permutation, for exchanging 15 and 14 (i.e., a single exchange) restores the natural order. On the other hand, the desired final configuration *is* the natural order, so that it requires no exchanges at all (or, if you prefer, two 'artificial' exchanges: 1 and 2, say, and then 1 and 2 again). In other words, the final configuration is *even*."

At this point Richter interrupted him, leaning forward, his voice filled with excitement: "I can now see what you're getting at, Mr. Davidson, and it's very clever. A permissible move of the game is actually *an exchange*: the number 16 (i.e., the empty square) and some other number (i.e., some other square) swap places. And so, after each single move, the new permutation on the board is odd if the previous one was even, and it is even if the previous one was odd. Therefore, to go from the initial configuration (odd) to the final one (even), if at all possible, must necessarily take an *odd* number of moves. That's your fact number 2, and the proof is complete, because the number of moves made by a player who had solved the puzzle would be both even and odd, a number that doesn't exist!"

Richter had leaped to his feet and joined Jule, who was still sitting at the computer desk. He put his right hand on Jule's shoulder, looked at him directly in the eyes and said, with a solemn expression on his face: "I have a very exciting and lucrative proposition for you, Mr. Davidson. I do hope you will accept it."

Only much later, curious about the history of the Fifteen Puzzle, did Jule learn the end of the story. It was obvious that Loyd had intentionally designed the puzzle so that it could not be solved. He had offered the $1,000 prize as an incentive for people to buy his game, well aware that he would never have to pay out the money. But despite the success of his invention, he never patented it in the United States. According to one source, in order to obtain a patent he had to submit a "working model." But when the patent office official learned from Loyd that the puzzle had no solution, he told him that in such case there couldn't be a working model, and without one, there could be no patent.

Chapter 4
A TRIP TO LONDON

As soon as he opened his eyes and had a look at his bedside clock on that Thursday morning in the late fall of 1997, Elmer James Galway had the odd feeling that his day was not going to be an ordinary one. To begin with, the sixty-three-year-old professor of classical history at Oxford University had slept late. This was not due to some alarm clock malfunction but rather to a dog failure. Slipper, his golden retriever—so named because of his predilection as a puppy for that particular object—had not climbed on the professor's bed and gently woken him up at 6 o'clock, as the dog dutifully did every morning, weekends included.

It was already 7:45 when Galway hurriedly got out of bed. As he headed for the kitchen to prepare his first cup of tea of the day, he bumped into an equally hurried Slipper coming in the opposite direction. "You're late, old boy, and so am I; we have no time for your morning walk. And it's all your fault," he said, pointing his forefinger at the dog and pretending to be annoyed with him. He received a loud bark in return.

The telephone rang before he could reach the kitchen. He took the call in his study, while a restless Slipper kept trying to get his attention.

"Elmer? It's David. I left you a message yesterday." David Green was the founder and owner of one of the leading companies of antiquarian book dealers in Britain: David Green Rare Books & Manuscripts Limited, and from time to time he sought Galway's expert advice on the origin or authenticity of some ancient manuscript. He was calling from his office on the top floor of the company's main shop, a three-story building in London's Mayfair district.

"I'm sorry, David, I got home late last night and didn't check my messages. What's up?"

"I'd very much appreciate your opinion on an item that has been offered to us. It's a rather urgent matter. Any chance you could come over?"

"Today?"

"If at all possible . . ." and, to make his request more attractive, he added: "It's something that might interest you, too."

"Hold on a second. Let me see . . . I know I'm already running late for the library committee meeting."

There was a short silence while Galway looked up his appointment book—he kept one at the college and another one at home, the two versions not always agreeing with each other. "Oh dear, I have an overseas visitor this afternoon. But I think I'll be able to manage if I . . ."

"That'll be great!" Green enthusiastically cut him off. "Any time today would be just fine. Thank you, Elmer; I really appreciate your help on this one." And he hung up before the professor had a chance to confirm that he would indeed be making the trip to London.

There were some arrangements to be made first. He could not miss the library committee meeting where substantial cuts to the scholarly journals subscription budget would be discussed and, without some staunch opposition, also approved. The meeting would likely take up the best part of the morning, but he could certainly catch the 12:45 train and be in London by 2:00. He avoided as much as possible driving to the city, which he considered an unnecessary hassle, preferring instead the convenience of traveling by train even if the quality and reliability of the service was no longer what it used to be.

It was too late to cancel his afternoon appointment. He would ask Bradley Johnston, his former student and now a junior faculty member, to look after his overseas visitor, a lady from some Canadian museum. She is probably interested in a guided tour of the Ashmolean's reserve collections, which are not open to the general public, he thought, and Bradley could perfectly well do it in my place. Funded in

1683 and part of Oxford University since the middle of the nineteenth century, the Ashmolean is Britain's oldest public museum. Among its treasures of art and archaeology there are major pieces of Greek and Roman sculptures, such as the Apollo from Olympia and the Prima Porta Augustus.

When Galway finally took a seat in the 12:45 direct train from Oxford to London Paddington station, he was very pleased with himself. The meeting had been less turbulent than he had anticipated and, more important, he had single-handedly prevented a drastic reduction in the department of classics funds for the purchase of learned journals. Some idiot from financial services had tried to justify the cuts by arguing that most of the information was available on the Internet for free anyway. The poor fellow obviously had no idea of how scholarly research was carried out.

But in reality, Galway was well aware that an increasing number of academic journals now published electronic versions that could be accessed at very low cost from practically anywhere at any time—and not only at the library during opening hours. Although he would have been reluctant to admit it, he mistrusted scholarly articles in electronic form and preferred the definitiveness of the printed version. Unlike its electronic counterpart, the latter could not be altered at will with a few keystrokes or clicks of the mouse.

"A compact disc might contain the equivalent of a small library," his defense of the printed word would start, "but you are a slave to the technology needed to read it. The ancient texts, on the other hand, written on traditional supports—clay tablets, papyrus scrolls, parchment, or paper—require no such intermediaries between text and reader. Without those 'hard copies' from the past we would know next to nothing of antiquity; and we'd better make sure that some hard copies of our own age are still around for the benefit of fortieth-century scholars."

Galway loved books as objects in themselves, and not only because of the information they contained. Reading a text from a computer screen was no match for the pleasure he derived from handling the bound stack of sheets, delicately turning a page or merely feeling the texture of the paper. He never missed an opportunity to browse through the shelves in Green's rare book store, located in the first floor of the Brook Street building. The place had the musty smell of dust and old paper typical of second-hand bookshops. Occasionally he would also visit the second floor, where the most valuable books and manuscripts were kept under special temperature and humidity conditions.

During one of these visits he had held in his hands a copy of the first printed translation into Latin of Euclid's *Elements*, dating from 1482. Surpassed in reputation only by Pythagoras and Archimedes, Euclid of Alexandria was one of the most prominent mathematicians of ancient Greece. He received his earlier training during the third century BC in Athens from the pupils of Plato, and later taught and founded a school in Alexandria. In his most famous work, the *Elements*, he developed geometry and the theory of numbers in a systematic way by logical deduction from certain basic assumptions or postulates, "bringing to irrefutable demonstration the things that had only been loosely proved by his predecessors," as the fifth-century Neoplatonist thinker Proclus put it. Through its countless editions and translations, Euclid's masterpiece became the most successful and influential mathematical textbook of all time.

The 1482 Latin edition up for sale was based on a medieval translation from Arabic by the English monk Adelard of Bath. Apart from the mathematical significance of the work, this particular edition inaugurated an era in the history of printing by featuring for the first time diagrams and geometrical forms cast in blocks of metal—a feat that the Venetian printer claimed no one before his time had been able to perform. The attractive, hand-colored copy was in excellent condition and had an early eighteenth-century inscription attesting its provenance from the library of a Jesuit convent.

Galway had opened the book at a random page and had translated part of the Latin text in his head. Without really trying to understand the mathematics, he had surmised that the author was discussing the prime numbers, that is, those positive integers that cannot be decomposed as a product of two smaller ones (the page actually contained Euclid's proof that there are infinitely many prime numbers*). When he inquired about the price, merely out of curiosity, he was

---

*See Appendix 2.

informed that "it is currently priced at 300,000 pounds"—about 450,000 dollars.

Over the years, Galway had built a small collection of ancient books and manuscripts, a few of them quite rare although none nearly as highly prized as the ancient copy of the *Elements*. But the pieces he treasured most were four Egyptian papyri scrolls given to him by his father. Galway junior had always felt uneasy about the papyri and would only show them or mention their existence to his closest friends. He suspected his father of having procured the scrolls by less than legitimate means during one of his field trips and so did not quite consider himself their rightful owner. They might be worth a small fortune, but Elmer had never dared to have the papyri appraised for fear of being questioned about their origin. Museums and reputable dealers in most countries now had a strict policy of not acquiring material that had left its country of origin illegally.

Times have changed, he would try to reassure himself; ethical standards in his father's time regarding artifacts discovered in the course of diggings or expeditions to remote corners of the world were much more relaxed than they are today. Take the nineteenth-century Italian explorer and one-time circus strongman Giovanni Belzoni, for example. One of the first Europeans to enter the Egyptian temples and pyramids, he shipped large numbers of Egyptian antiquities back to the British Museum in London, notably the colossal bust of Ramesses II. In his 1820 best-selling book *Narrative of the Operations and Recent Discoveries within the Pyramids, Temples, Tombs and Excavations in Egypt and Nubia*, he gives some idea of the archaeological practices of the time: "Thus I proceeded from one cave to another, all full of mummies piled in various ways, some standing, some lying, some piled on their heads. The purpose of my researches was to rob the Egyptians of their papyri; of which I found a few hidden in their breasts, under their arms, above their knees, or on the legs, and covered by the numerous folds of cloth that envelop the mummy." At least Belzoni had the courtesy of leaving some papyri for future archaeologists—such as his father?—to help themselves, thought Galway sarcastically.

He had never mentioned to his father his misgivings about the scrolls—nor, for that matter, about the vast assortment of other artifacts, from coins to sarcophagi, that Sir Ernest had brought back from excavation sites over the years—partly due to the fact that the exuberant father and his reserved offspring did not get along all that well, but he did seek the advice of his elder brother John.

"Nonsense, finders keepers," was the not-so-subtle response he received from him. And John went on: "Had it not been for the British and the Germans who dug 'em up, all that stuff would still be rotting in the caves or wherever it was they were found. You know better than I that before Schliemann got there, the Greek peasants were using those old stones so precious to you to build chicken coops. And how many thousands of ancient records and literary works did Egyptian peasants set on fire just to get a kick out of smelling burning papyrus? We did those people and humanity a favor, Elmer." Heinrich Schliemann was one of the pioneers of modern archaeology. German-born and later an American citizen, he conducted extensive diggings in the late nineteenth century in Greece leading to the discovery of the ruins of Troy, the shaft graves with their immense treasure of gold, silver, and ivory objects at Mycenae, and the palace of Minos at Knossos, on the island of Crete.

John Arthur Galway had the flamboyant personality and the outspoken manners of the successful businessman he actually was. He had a Master's degree in geology but never intended to become a scientist. Fresh from university, he founded an import-export company dealing in gems—emeralds in particular but also rubies and other precious stones. He was also a junior partner in a multinational corporation mining for emeralds in Colombia.

John had his own idea of what his younger brother should do with the scrolls. "I know a chap in Bogotá who'd pay a pretty penny for those papyri of yours. No questions asked," he told Elmer. "You'd stop having qualms about them and on top of that you'd be set up with a jolly bundle for your retirement. What do you say?" Elmer politely replied that he would think about it, but what he was really thinking was that his brother had even fewer scruples than his father.

The train was pulling into London's Paddington station on time. At that early afternoon hour the place was already bustling with people coming and going, and it took Galway a good twenty minutes to get a taxi. He gave the driver the address of David Green Rare

Books & Manuscripts Limited. Traffic was heavy, and it took him another twenty minutes to reach his destination.

The first floor of the building was open to the public, whereas access to the upper floors was by appointment only. Galway entered the shop and went up the stairs to a landing with a chair on one side of a closed door and a small table on the other. The sign on the door read "Access restricted. Please ring for service." He rang the intercom buzzer, identified himself, was let in, and continued up the stairs. When he got to the next and top floor, he walked down the narrow corridor toward the reception desk.

"Professor Galway, so good to see you!" It was Sandra, Green's secretary, greeting him in her usual merry mood. Sandra, a brunette of powerful build and still unmarried in her late forties, was totally committed to her job. Her warmth and good humor made clients and visitors alike feel welcome as soon as they came in. "Hello, Sandra; good to see you too. How's Beauty doing?" Galway was inquiring after Sandra's cat. "She's been unwell lately, the poor darling. The vet put her on a special diet. Seems she's been eating too much, just like her mistress!" She laughed. Galway smiled uncomfortably, unsure of what to say next.

"Elmer, so glad you could make it!" interrupted David Green coming out of his office. Saved by the bell, Galway thought. Green gripped the professor's right hand with both of his and shook it vigorously while he invited him to come in. And after a "No interruptions, Sandra; thank you" to his secretary, the two men disappeared behind the closed door of Green's office.

Chapter 5

A LETTER FROM THE PAST

Elmer Galway and David Green were sitting facing each other in a cluttered office on the third floor of Green's rare book shop in West London. They were separated by a glass-top table; a single object, wrapped in a piece of cloth, sat on it. Outside it was already dark at three o'clock in the afternoon and a fine drizzle was falling. It was equally dark inside the room, except for a floor lamp that spilled its cone of yellowish light on the table.

They had been together for fifteen minutes, and up to that point Green had been doing most of the talking. He had told the professor all about Señor Alfonso Lopez de Burgos, a Spaniard who was in London on a business trip and who had come to the shop the day before, bringing with him an ancient book that he wished to have appraised with the intention of selling it. The book in question was the object lying on the table, and Green would very much appreciate Galway's expert opinion as to its origin and authenticity, as well as its possible historical value. Time was of the essence because Sr. Lopez de Burgos was leaving the next day for Hamburg, where he also intended to have the book evaluated. The Spaniard obviously believed that he had in his possession a valuable item and was shopping around in search of the highest bidder. No, Sr. de Burgos was not one of his regular clients, Green informed Galway, in answer to his question.

Leaning forward, Galway picked up the covered book from the table and carefully removed the wrapping cloth. He was wearing white cotton gloves, a standard procedure for handling ancient objects in order to prevent further damage and at the same time protect the handler from mold and other potentially toxic substances. He deposited the piece of cloth on the table and began to examine the manuscript.

The "book" consisted of a dozen or so loosely bound six-by-nine-inch parchment sheets, with a front cover but with no back cover or title page. Its condition was fairly good, considering that it had probably survived a fire, as evidenced by the charred edges of the pages. It was written entirely in ancient Arabic, a language with which Galway had some familiarity from his own work on classical history. From the eighth to the thirteenth century, while most of Western Europe was decimated by epidemics, famine, and wars, Arab scientists and philosophers rescued Antiquity's cultural heritage from oblivion thanks to their translations into Arabic of the works of Plato, Aristotle, Euclid, Archimedes, Ptolemy, Plotinus, and many other great thinkers. These Arabic versions—covering a wide range of subjects and genres, from philosophical discourses and literary creations to scientific treatises and technical manuals—were often augmented with the translator's own contribution in the form of commentaries or additions. While the great majority of the original Greek sources written on papyrus were lost, most of their Arabic translations survived and provided modern scholars with an invaluable tool for the study of the golden age of Greek culture and civilization.

"And how did Sr. de Burgos come into possession of this manuscript?" asked Galway without lifting his eyes from the book.

"He mentioned that it had been in the family for many years but couldn't give any specifics. He comes from the city of Córdoba, where some of his ancestors lived as far back as the Middle Ages, when the city was under Moorish rule."

"Córdoba was once the capital of Moorish Spain and the center of an independent caliphate." It was the history professor speaking, this time taking his eyes off the book to look at Green. "The greatest Arabian philosopher in the West and famous commentator of Aristotle, Ibn-Roshd, better known as Averroes, was born in Córdoba. The city reached the peak of its splendor in the middle of the tenth century. At the time it was a leading intellectual center, and it possessed one of the largest and richest libraries in Europe."

"Are you suggesting that this Arabic manuscript may actually come from Moorish Córdoba?" asked Green without much enthusiasm. He had examined the book and tentatively dated it not older than the sixteenth century but he had little idea of its content; all he knew was what Sr. de Burgos had told him: that it was an Arabic translation of an ancient Greek manuscript, a tale or an epic of some sort. "From what the client told me," he went on, ignoring the fact that Galway had not answered his previous question, "the book might have found its way into the Lopez de Burgos family through one of his ancestors in the Middle Ages. But I don't think so; the parchment doesn't look to me nearly old enough for that."

But Galway was no longer listening to Green. He had begun deciphering the Arabic text with the help of a magnifying glass. The handwriting was relatively easy to read, but the black ink, most likely iron gall, had eaten away the parchment at certain places. Green sensed that the professor was now focusing all his attention on the manuscript and decided not to disturb his concentration. He just sat there, deep in his own thoughts, and from time to time glanced at Galway, who was showing increasing excitement as he skimmed through the book, apparently only trying to get the general sense of the text. The telephone on Green's desk rang twice before Sandra answered the call from the reception desk, but neither man seemed to take notice.

Almost fifteen minutes went by before Galway at last closed the book, slowly put the magnifying glass back in his pocket, stared at Green from over the rim of his reading glasses, and in a deliberately calm voice said: "If this is what I think it is, you are being given the chance to acquire a most extraordinary document."

Galway's words clearly took the antiquarian bookseller by surprise. He appeared more bewildered than pleased by what he had heard and could only manage a weak "Why . . . what do you mean?" in response.

"This manuscript is most likely a letter," Galway began to explain, "and what its author tells is a story all right but not a mythological epic, as a superficial reading of the text might suggest. To the trained eye of someone with the proper knowledge of pre-Socratic Greek history and capable of some minor extrapolations to fill the gaps, however, what I'm holding in my hands is an eyewitness account of the death of the great philosopher and mathematician Pythagoras and the end of his inner circle of followers known as the early Pythagoreans."

Green remained silent, as if expecting the professor to go on. "To be sure, Pythagoras' name does not appear anywhere in the text," Galway resumed, "but this is consistent with the practice among his disciples who, out of reverence, never pronounced the philosopher's name. Moreover, the fact that Pythagoras is not named suggests that the author is one of his followers, and also that his account was only intended for internal consumption, so to speak, which is in line with the shroud of secrecy that the Pythagoreans maintained on matters concerning the fellowship." He took off his reading glasses before proceeding. "Naturally, this is only an Arabic translation of the Greek manuscript, dating probably from the thirteenth century—or perhaps even a copy of the translation. Whatever the case, if the author is telling the truth and if the document is authentic—two big 'ifs,' I grant you—we now know that Pythagoras perished in the fire set by an angry mob to the house where he and his followers were gathered in the city of Croton, in what is now southern Italy. The reasons for them being there and the name of the instigator of the attack are given, and I'm sure that a thorough analysis of the text—I have only managed to decipher the gist of it—will reveal many other details, until now unknown to historians, regarding the circumstances surrounding the philosopher's death and the fate of his surviving disciples." He gently tapped the ancient parchment with his cotton-covered fingertips as he said: "It would be impossible to exaggerate the historical significance of this document. The oldest extant accounts of this tragic event, believed to have taken place at around 500 BC, date from the middle of the fourth century AD—almost nine hundred years after the fact—and some versions have Pythagoras surviving the fire and fleeing to Metapontum where he supposedly died many years later."

"What do you want me to do, then?" asked Green, already recovered from the initial shock.

"For the moment, it is essential that you buy some time. Tell Sr. de Burgos that you're interested in the book but that you need more time to study it and to have some tests performed on the parchment and the ink; but don't give him reason to believe that his manuscript may be a historical bombshell lest the asking price should be multiplied twentyfold. In the meantime, I'll prepare a complete translation as soon as possible. You've got a copy on paper, haven't you?"

"Sure," said Green, nodding. To avoid exposing the parchment to the light and heat generated by a photocopying machine, he had photographed it using a digital camera that required no flash, and had printed a computer-enhanced copy of each of the twenty-two pages that made up the book. He then remembered having noticed something peculiar as he photographed the last page. The back cover was missing, but that was not unusual in very old books. It was something else. "Are any pages missing?" he put the question to Galway with a sudden excitement in his voice.

The professor had already begun wrapping the book back in its enveloping linen cloth. "It's possible; the adhesives on the spine binding are quite dry and as a result some pages may have become loose and fallen off, but you know that better than I," Galway said, a bit puzzled by the question.

"I mean, are there any pages missing *at the end*?" insisted Green. "It's hard to tell. Since the protective back cover is missing, the last page is covered with various residues and accretions, and it's practically illegible," said Galway as he removed the cloth once again and began examining the back of the manuscript. Green, who by now was standing beside him, pointed his finger at the spine. "There, have a look at the edge of the spine with your magnifying glass." Galway did as instructed. "I'll be damned!" he exclaimed, "It's been cut clean!"

It did not take them long to piece together a likely explanation for their discovery. The original book had been separated into two, more slender, "half-books" by cutting the spine lengthwise. That might explain the absence of a back cover, which presumably had remained attached to the second half. Then, the edge of the remaining spine had been roughed up and dyed to make it appear worn out and to conceal the fact that it had been cut; finally, the last page had been scraped off and soiled to render it illegible. They were convinced that a closer examination of the parchment would confirm their theory. As to the reasons for the "surgical" operation, they could only speculate.

Green offered his conjecture: "I suspect that Sr. Lopez de Burgos believes that two ancient half-books are worth more than a whole one—which is probably true, I'm afraid."

"If so, this is a case of the whole being *less* than the sum of its parts," said Galway with a smile; and, more seriously, he added: "If our guess is correct, what happened to the other half of the document?"

"It might have already been sold. . . ."

"True, but it might not have been recognized for what it really is, especially without the first part. I wonder what else the author had to say, since what we've got seems to contain the main episode of the story. We'll probably know more after I complete the translation. I'll get down to it right away and hopefully shall be able to send it to you by Monday."

"That would be great, but my more immediate concern is Sr. de Burgos. What if he wants an estimate by tomorrow morning, when he is supposed to come to the shop? He may insist on taking the book with him to Hamburg."

"You'll have no choice but to stick to your line: you're interested in his manuscript but you need more time before you can put a price tag on it. Let him decide what to do next. I'm afraid it's no longer in your hands."

When Elmer Galway arrived home, well past 10 o'clock that night, he was too excited to go to sleep and decided to start working on the translation of the manuscript right away. He changed, prepared himself a fresh pot of tea, and sat down at his desk, with Slipper sprawled on the faded Oriental carpet at his feet.

Before getting down to work, he checked his voice mail. There was a message from Bradley Johnston, who wanted to see him about his Canadian visitor but would be out of town until Monday; there was also a call from the veterinarian's office reminding him that Slipper was due for his annual shots; and there was the following recorded message from David Green: "*Elmer, it's David. Sr. de Burgos called. Someone broke into his hotel room and he no longer wishes to take the book with him to Hamburg. He wants me to keep it in my safe until he comes back next week. Thought you'd be glad to hear we have a few more days to complete the evaluation. Looking forward to receiving the translation.*"

Chapter 6
FOUND AND LOST

M onday came and went and Green had not yet heard from Galway. The translation may be taking the professor longer than
he had anticipated, he thought. But he didn't mind, for he had his
hands full with the preparation of the new catalogue.

On Tuesday morning, almost one week after he had first called,
Alfonso Lopez de Burgos was back at Green's shop. In the intervening
time his hotel room had been broken into and he had spent three days
in Hamburg on business. He was in a rather somber mood, contrast-
ing with the expansive customer that had walked into the shop six
days earlier with an ancient book in his briefcase.

As he had insisted on seeing Green in private, they were in a small
room next to Green's office that was used on such occasions. The
Spaniard was over six-feet-three and built like a weightlifter, dark-
eyed, with a balding head and an immaculately trimmed thin mous-
tache. He was wearing a gray pinstripe business suit and shiny black
shoes. Green invited him to sit down in a large armchair next to the
window, while he took a seat on the couch opposite him. A couple of
chairs, a shaky table on which sat a coffeemaker, a file cabinet, and a
low bookcase completed the furnishings in the room, which Green
used mostly as a place to read and relax.

The book dealer had prepared a plan of attack. He would confront
de Burgos with the mutilated spine and ask for an explanation
without directly accusing him of wrongdoing. He would then argue
that the loss of the last part negatively affected the value of the

manuscript—the opening salvo in the bargaining battle that would certainly follow.

But Sr. de Burgos' unexpected move preempted Green's well-planned offensive: he had decided to come clean about the missing half of the book. There was one condition attached, though: Green had to promise not to divulge what he was about to hear. "You can count on my complete discretion," Green assured the Spaniard, but he quickly added: "Unless, of course, there is some criminal activity involved."

"*No, no, nada criminal,* nothing criminal," said de Burgos emphatically. And then he asked: "Are you a religious man, Mr. Green?" The question took the book dealer by surprise. "Well, yes . . . in a sense I suppose I am." This half-hearted admission seemed to reassure his client, for he immediately began to talk. Here is the story he told Green.

On September 27, 1997, an earthquake hit the region of Assisi, in central Italy, badly damaging the magnificent thirteenth-century basilica and convent of Saint Francis. Somewhere in the bowels of the lower church, the crumbling of a wall revealed a hidden chamber, where various relics and religious objects, together with old church records and a small collection of ancient books, were found. Most of the documents were in very poor condition, but a few items had been relatively well preserved.

Fra Benedetto, the convent's chief archivist and librarian, believed that the objects had been deliberately walled up in order to protect them, probably back in the fourteenth or fifteenth century. In those times, Assisi and the neighboring Perugia were bitter enemies, and the Perugians had on more than one occasion captured and sacked their rival city, destroying and burning many of its treasures.

Benedetto then decided to sell some of the ancient books to raise money for the reconstruction of the basilica and the restoration of its precious frescoes, the work of the medieval Italian masters Giotto and Cimabue. Only those manuscripts that he deemed of no historical interest for the Church or the Franciscan order were to be put up for sale. The entire operation was to be conducted privately and in secret, for the friar was acting on his own, without having sought permission from the Order's higher authorities. Benedetto felt that his action was inspired by the founder of the Order himself. According to a story, Saint Francis was praying one day at the neglected church of Saint Damian, in Assisi, when he heard a voice from the crucifix summon him: "Francis, go repair my house that thou seest is all in ruins." The impulsive young Francis then secretly sold some silks from his father's warehouse to finance the repair project and so fulfill the Lord's wishes.

But there was an even more important reason for not having an open sale: the ownership question. The fact that the artifacts and documents had been found inside the church was not automatic proof that they belonged to the Fraternity. Trying to establish their ownership through the proper channels could mean having to wait a long time. It was not until 1929 that the Holy See was officially recognized as the owner of the centuries-old Franciscan Archives through a concordat—a pact concluded between the Pope and the secular authority. The chain of ownership prior to that date was not clear.

Fra Benedetto really had no choice but to embark on a covert operation. The game was worth the candle—or, in this case, a new dome for the basilica was well worth some circumventing of proper procedure. He prayed to the Lord, asking for His understanding if not His forgiveness, and went ahead with his scheme.

Benedetto confided his plan to Fra Ignacio, his longtime friend and current treasurer of the Order. Fra Ignacio was a Spaniard from an old aristocratic Spanish family; he was also Sr. Lopez de Burgos' elder brother.

Ignacio went along with Benedetto's plot and proposed to sell the ancient books with the help of his brother Alfonso, a devout Catholic who traveled extensively on business and in whom he had complete trust. The idea was for Alfonso to offer the books for sale one at a time and at different places so as not to arouse suspicion, and to claim that the document had been acquired by one of his ancestors so long ago that no record of ownership was available. The reputation of the Lopez de Burgos family, which for more than seven centuries had given its country generals, ministers, and ambassadors and the Roman Catholic Church archbishops and cardinals, would help to establish Alfonso's credibility—or so they thought.

The plan had not worked as smoothly as they had imagined. Alfonso had indeed managed to sell four of the five valuable books.

But what he didn't tell Green was that, except for one shop in Madrid, respected antique dealers were not convinced by his "it-was-in-the-family" story and, sensing some foul play, had declined to buy. The Spaniard had then been forced to deal through middlemen of questionable reputation who were buying in the black market on behalf of anonymous collectors and who would offer only a fraction of the estimated price of the manuscript.

"And how much money have all these transactions brought in, if I may ask?" interrupted Green. He reckoned that the Pythagorean parchment in his custody could easily fetch three hundred thousand pounds. If the rest of the items were of comparable value, the total take could be close to a couple of million.

"I'm afraid I'm not at liberty to say," replied the Spaniard, and he quickly resumed his story as if to fend off any further questioning.

The fifth and last book was a special case. The last eight pages were written in a different language—Greek, he thought—and contained what he took to be some geometric illustrations or artwork. Just as Green had conjectured earlier, de Burgos had separated them from the rest of the book by cutting the spine lengthwise, thus obtaining another "book," this one with samples of medieval art, that he hoped could be sold separately and increase his total take.

What de Burgos had done was not unusual. It was well known in the trade that some careless or unscrupulous individuals would not hesitate to tear off beautiful pictures or illuminations from valuable books, often with disastrous results for the integrity of the original document, expecting to reap a bigger gain by selling them separately.

Knowing what he did about the contents of the Pythagorean manuscript, Green was puzzled by the fact that a historical account should end with a series of illustrations. Were these mere decorations designed to render the document more attractive, or were they an integral part of the story being told? And why the change of language, from Arabic to Greek? It was not unusual for Roman volumes to include commentaries in Latin of a Greek text for the benefit of those who could not read Greek. Was the Arabic text a commentary of the Greek here too? He would have to ask Galway about these things, but at present his more immediate concern was the fate of the missing pages.

"And where is that second 'book' now?" he asked.

There was a long silence before an embarrassed de Burgos answered, looking away from Green: "I don't know."

"What do you mean you don't know?"

"*Me lo han robado.* Someone took it from my hotel room. Last Thursday afternoon, when I got back to my room, I noticed that my suitcase had been opened and searched. I don't usually keep any valuables in it; all documents related to my trade—I'm in the security systems business—I carry with me in my briefcase or my laptop at all times." He told Green that he had left the mutilated book in his suitcase, inconspicuously buried in a pile of brochures, magazines and various other papers, thinking it would be safe. But it was gone. "*Desapareció.* And only someone who knew what it was—and what it might be worth—could have taken it."

"Did you report the theft to the police?"

"No, I didn't, not even to the hotel management, and you can understand why: I would have had some explaining to do. Besides, I don't think the police—or anyone else, for that matter—can help me get the parchment back. It's gone forever, Mr. Green, and I dread the moment when I break the bad news to my brother. I'm so ashamed. I should never have tampered with the book, but I did and was punished for my greed."

Green wished he could believe the Spaniard's story. But did it really matter? Occasionally, he had asked vendors to sign a document declaring that a book is offered "free from any legal encumbrance," that is, assuring that they own it and have a right to sell it. This provided him with a degree of legal protection. But the man had admitted that the book was not his. On the other hand, if he was telling the truth the Franciscan Order could lay claim to the volume, presumably with success, giving them the right to sell it. Then, an affidavit from either Fra Benedetto or Fra Ignacio as the Order's representative would suffice, and it could be kept confidential if they so wished.

There was still another question. Ordinarily, Green would try to pay as little as possible when purchasing a book and then make a profit, as large as possible, when he sold it. That was simply how the laws of the market worked, how a profitable business was supposed to be run. But if the story he had been told was true, this was no ordinary commercial transaction. The Catholic Church was involved, and the money they expected to collect would go toward a worthy cause. How could he then offer de Burgos, who was not aware of the Pythagorean manuscript's true nature, a sum well below what he was almost certain to get for it? How could he so shamelessly take advantage of the situation? He decided he couldn't.

"Here's what I propose, Sr. de Burgos." The Spaniard was all ears: maybe there was an honorable way out of his predicament. He had done the right thing by trusting in Green, he thought.

"I would need a written statement from your brother or Fra Benedetto," Green resumed, "duly authenticated by a solicitor, relating the circumstances surrounding the book's discovery and laying claim to it. This document should also give you power of attorney to sell it."

"*Sí, sí, es posible.* Yes, I understand. I think it can be done."

Green then explained that not only did he not want to be involved in a crime but he was protecting himself against a commercially unsound operation. "For it is not unheard of for rightful owners to appear some time down the road to claim ownership," he explained. "Then, a chain of sales between owners and dealers would have to be unraveled, legally and financially, resulting in considerable losses to all parties."

"So you are willing to buy my book. How much do you offer me for it, then?" Sr. de Burgos was anxious to conclude an official sale and receive the fair market price of the book—or what was left of it.

"Yes, of course I'm interested, but I don't wish to buy your book."

"I don't understand. . . ."

"I'm still waiting for an expert report, but if the manuscript is authentic—and I believe it is; very old parchments are extremely difficult to fake—I propose to sell it at an auction we will be holding in January. That way you'll maximize your take, and Saint Francis will have his basilica restored sooner."

De Burgos half raised from his seat and grasped Green's arm with both hands as he said, "Thank you, thank you very much, Mr. Green." An expression of relief filled his big round face.

"There's one more thing I need from you, though." Something in Green's voice seemed to suggest that there was a problem. The Spaniard retreated back into his chair.

"I cannot sell publicly an important artifact that came to the United Kingdom without Italian and European Union export licenses," said Green. "Much as I would like to help you, I'm not willing to risk my reputation to do it."

His earlier excitement rapidly turning into disappointment, de Burgos stared at Green, who was pondering how to best formulate his next question—or rather how to disguise a delicate request as a question.

"Is there any way your brother could come into possession"—Green hesitated—"Could he somehow obtain the necessary documents? I mean . . ."

"I know what you mean," interrupted de Burgos with an understanding grin. "I'm sure it can be done; the Order has friends in high places."

"Very well, then. As soon as you have the affidavit and the rest of the documents, fax them to me and I'll make the necessary arrangements for the auction." And, somewhat ill at ease, he added: "My fee for the transaction will be 5 percent of the selling price, which is below the standard commission of 7 percent."

When Alfonso Lopez de Burgos left David Green Rare Books & Manuscripts Limited's main shop late in the afternoon, he was again a happy man. Surely, there were still some details to iron out and he had yet to tell his brother about the theft, but thanks to Green he now saw things differently. His glass was no longer half-empty, it was half-full: he considered himself lucky to have lost only one half of the valuable manuscript and had high hopes of getting a considerable sum for the other half. True, he should have been more careful and deposited it in the hotel's safe, but how could he have suspected that someone would find it in his suitcase and take it away? He had instinctively assumed that the intrusion was connected with his trade—a competitor trying to steal some industrial secret. In the security systems world, espionage was a constant threat and he was used to it. He always carried all his business material with him, and the most sensitive documents were encrypted. But then another possibility had dawned on him: Could it be that whoever took the ancient parchment broke into his room expressly looking for it? It didn't seem possible. Nobody knew

he was carrying a valuable book, besides his brother and Fra Benedetto, of course. Just the same, as a precaution, he had decided that
the remaining half of the book would be safer in Green's shop than
with him on his trip to Hamburg.

Back in his office, David Green was asking himself similar questions: Was the intruder after de Burgos' book? Was someone else on
the trail of the Pythagorean manuscript? Why didn't he ask de Burgos
about this? I need to talk to Galway, he thought.

He had entered his office through a connecting door and was sitting at his desk, pondering his conversation with the Spaniard. He
pushed a button on the intercom. "Sandra, would you please get Professor Galway on the phone? If you leave a message, tell him it's rather
urgent. Thank you."

He did not have to wait long. "Mr. Green?" There was an edge to
Sandra's voice. "I've just talked to his assistant. Professor Galway's father passed away and he's in Cardiff for the funeral."

Chapter 7
A DEATH IN THE FAMILY

Sir Ernest James Galway died peacefully in his sleep on November 28, 1997, the eve of his ninety-fourth birthday. The famous archaeologist lived alone in a semi-detached house with a small garden at the back, situated not far from the center of Cardiff. A housekeeper, who came on Fridays, discovered the old man's body in his bed and communicated the sad news to his sons.

By most standards, Ernest Galway's life had been a full and happy one. Born in London in 1903, he was too young to serve his country in the Great War and too old for combat duty during World War II. He had traveled extensively and enjoyed a long and brilliant career as an archaeologist and explorer, during which he had received a multitude of prizes and distinctions. His married life had been less successful, but he had stayed on good terms with his former wife after their divorce, and had never lost contact with his two sons, John and Elmer. These had now come to pay him a final visit.

On Tuesday, after the funeral, Elmer and John went through their father's belongings. The old man had long given away to museums and universities most of the ancient artifacts and other objects of archaeological value he had collected over the years. Among those he had kept were several brightly colored African masks and shields which decorated one of the walls of his study. On his desk and in a filing cabinet they found records of his numerous trips and expeditions: photographs, notes, letters, drawings, maps, contracts, receipts, and so forth. They knew that their father had been working on his memoirs, a project he had undertaken a few years ago. He used an old Macintosh LC to type the manuscript, a printed version of which, identified as "Chapters 1–14, 3rd draft," they found in one of the desk's drawers.

Ernest Galway's attorney, Robert Harris, a friend of the family, was the executor. The terms of the will were straightforward: John and Elmer inherited in equal parts all his possessions, except for a sum of 5,000 pounds, kept in a special account, which was to be donated to the South Wales Archaeological Society. There was one special clause: Should he die before his memoirs had been published, all related documents, preliminary drafts, computer files, etc., were to be entrusted to Elmer, who could then dispose of them as he wished. Before leaving Cardiff, Elmer made arrangements to have these documents and his father's computer sent to Oxford.

In the afternoon, Elmer called Green and learned about Sr. de Burgos' story and the upcoming auction. For his part, Galway told Green that the Pythagorean book appeared to be the Arabic translation of a letter from one disciple of Pythagoras to another. "As far as I can tell," he said, "after performing a preliminary philological analysis— comparing vocabulary, style, abbreviations, and so on with reliable sources—the Arabic text was written around the twelfth–thirteenth century."

"That checks," interrupted Green. "I just got the results from the laboratory tests on the parchment and the ink: with 90 percent certainty, they concluded that it dates from sometime between the early twelfth and the middle thirteenth century."

"Excellent," said Galway, and then: "Also, I believe that the translation is probably a copy, for there are some errors that a translator wouldn't have made but could easily occur when a scribe is merely copying in a mechanical fashion—this doesn't detract from the value of the book, though. Besides relating the circumstances of Pythagoras' death, the author also claims to be in possession of a document written by Pythagoras himself, which, he writes, 'must be protected at all costs.' This is another historical shocker, for all authorities on ancient Greece are unanimous: Pythagoras did not leave behind any writings. Only after the philosopher's death, fearing that their master's teachings might be lost forever, did some of his disciples write a collection of abstracts and commentaries. So, if a manuscript in Pythagoras' own hand really existed—most likely a papyrus scroll—it would be a document of exceptional historical value. Moreover, one is left to wonder what might have prompted him to write it, for he refrained from putting his discoveries in writing lest they should fall into the hands of the impure. What could have been so important to warrant abandoning his principle against the written word?"

Galway believed that the pages missing from de Burgos' book may not be mere embellishments and might throw some light on the fate of Pythagoras' scroll.

"Did you ask de Burgos if he had photocopied the book?"

"No, as a matter of fact I didn't. But I presume he would have told me if he had."

"Why don't you ask him, just in case? I'd be surprised if the Franciscans didn't make a copy of it for their archives. They are compulsive record-keepers."

Galway expected to finish the translation by Thursday. He would then e-mail it to Green, together with a short summary to accompany the Notice of Auction.

"As you would understand," he told Green, "I wish to retain copyright to the translation and eventually submit an annotated version of it for publication in some prestigious journal. But I'll wait to see if we can obtain the missing pages, or photocopies of them. In the meantime, the text I'll send you is 'for your eyes only,' as they say in that James Bond movie."

Green then asked Galway whether the Ashmolean Museum might be interested in purchasing the Pythagorean book. "Perhaps, but I doubt they could outbid all those wealthy private collectors."

On Wednesday, Galway was back at his office in Oxford and began to clear the backlog of e-mail and voice mail messages accumulated during his absence. He then asked Bradley Johnston to drop by—he had some news for him.

Johnston first told Galway about Irena Montryan's visit and her interest in manuscripts from Pythagoras' time for her exhibition. Talk about coincidence, thought Galway, and then he said: "Well, after you hear what I have discovered you'll be able to tell that young lady from

Canada that there's a chance she might get what she's looking for."
And, after a slight hesitation, he added: "But warn her not to get her
hopes too high."

The professor seemed to imply that Johnston was now in charge of
dealing with Irena's request, which suited his former student just fine,
happy to be given a reason to stay in touch with her.

When Galway got home that evening, he was greeted by an excited
Slipper jumping all over him. The poor animal had not yet completely
recovered from the four days he had spent in the kennel while his
master was away. "I know you missed me," he said to the dog, pinch-
ing both his cheeks and gently shaking his head. "I missed you too, old
boy."

He had difficulty falling asleep. Questions kept popping up in his
mind. Even if Pythagoras really did write something on a papyrus,
what were the chances that the scroll still existed 2,500 years later?
Rather slim, really, although much older Egyptian papyri had been
found in relatively good condition. "Must be protected at all costs,"
the author of the ancient letter had written. Perhaps extraordinary
precautions had been taken to preserve the papyrus from the ravages
of time—and from undesirable human hands. Would that mean that
the precious scroll was hidden somewhere, maybe so well hidden that
it would never be found? But if it existed at all, the scroll must have
been written for someone, for what would be the point of writing it if
no one could ever read it? Had Pythagoras intended it as some sort of
time capsule, to be discovered and its contents revealed at some point
in the future? And if so, when exactly and for what purpose?

Galway realized that there was only one way to find out: embarking
on a search for Pythagoras' manuscript, even if there was only a remote probability of success. As if reaching this conclusion had released the tension that was keeping him awake, he immediately fell
sound asleep.

PART II

AN EXTRAORDINARILY GIFTED MAN

Chapter 8

THE MISSION

In the age of Leucippus and Democritus, and even before them, lived those called Pythagoreans, who applied themselves to the study of mathematics and were the first to advance that science; and, penetrated with it, they fancied that the principles of mathematics were the principles of all things.

—Aristotle, *Metaphysics* A.5

Lysis woke up abruptly, his heart thumping, his forehead covered with sweat. He had been dreaming of fire—again. Only this time the dream had seemed so real he could have sworn he had felt the scorching heat on his face and heard the screams, people yelling as if in pain, crying for help. He was an optimist by nature, used to looking at the bright side of things, but lately he had become possessed with a feeling that those dreams forebode some imminent calamity. The time was the second year of the 71st Olympiad, or 495 BC.

Could my dreams be a warning from the gods? He wondered. It was a while since he had sacrificed to Hera, the queen of the Olympian gods. When he worshipped at her temple situated in a promontory overlooking the Ionian Sea near the city of Croton, in Magna Graecia, he would leave wheat and barley and cheese cakes as offerings. He knew that the cow was the animal especially sacred to the goddess, but the killing of animals was forbidden in the fraternity— the sect, as outsiders called it—as was the eating of animal flesh. As prescribed by the Master, he would enter the temple from the left and wearing a clean garment in which no one had slept, because sleep, just as black or brown, indicates slothfulness, while cleanliness is a sign of fairness and justice in reasoning.

He wished he could travel to Delphi to consult the oracle about his dreams, but right now he had to attend to a more pressing affair. The Master had sent for him concerning "an urgent matter of the utmost importance," as Zalmoxis, the Master's servant, had put it when he delivered the message the night before.

Lysis came from the city of Tarentum, situated on a rocky peninsula north of Croton. As a young man he had traveled to Croton to listen to a sage by the name of Pythagoras who had arrived in the city from abroad. Tall and graceful in speech and gesture and blessed by nature with unsurpassed intelligence, the well-traveled foreigner had made a strong impression on the Crotonians and rapidly won their esteem, to the point that the governing council of elders had invited him to give lectures also to the younger men and the women. Such was the man's reputation that people from far and wide, including magnates and kings, came to hear his eloquent discourses.

"Observe the crowd gathering at some public spectacle," Pythagoras would tell the audience, "and you will see men of all descriptions and views. One hastens to sell his wares for money and gain; another exhibits his bodily strength for renown; but the most liberal assemble to contemplate the landscape, the beautiful works of art, the specimens of valor, and the customary literary productions. So also in the present life men of manifold pursuits are assembled. Some are influenced by the desire for riches and luxury; others, by the love of power and dominion, or by an insane ambition for glory. But the purest and more genuine character is that of the man who devotes himself to the contemplation of the most beautiful things, and he may properly be called a philosopher. The survey of the whole heaven and of the stars that revolve therein is indeed beautiful, when we consider their order, which is derived from participation in the first intelligible essence. But that first essence is the nature of Number and numerical ratios that pervades everything, and according to which all those celestial bodies are arranged elegantly, and adorned fittingly. Similarly beautiful is devotion to erudition. The desire for something like this is philosophy."

He would also teach that the soul is immortal, and that after death it transmigrates into another human or mammalian animal body.

For this reason people must abstain from eating flesh, lest they unwittingly should devour their own parents, or children, or friends in some altered shape. Beans too were forbidden for various reasons, but especially because they were believed to be a temporary receptacle for the transmigration of souls, perhaps due to their resemblance to human testicles.

"Do not defile your bodies with impious food," he would urge the crowd. "We have corn; we have apples weighing down the branches, and grapes swelling on the vines. There are flavorsome herbs and vegetables that can be cooked and softened over the fire, and there is no lack of flowing milk or thyme-scented honey. The earth, prodigal of its wealth, offers us food that can be procured without slaughtering or bloodshed."

Pythagoras was a native of the island of Samos, in Ionia, a Greek colony on the western coast of the Aegean Sea. Around the middle of the sixth century BC, the once thriving and opulent Ionians had fallen under the yoke, first of the Kingdom of Lydia, then of Persia. At a time when the victorious days of Athens and Sparta were yet to come, the glory of the Hellenic spirit was kept alive by the Greeks who settled in southern Italy. Following the Persian conquest, many eminent Ionians migrated there, attracted by the freedom and prosperity of the Greek cities that dotted the lower part of the Italian boot. One of these newcomers was Pythagoras. Soon after arriving in Croton, he founded an ascetic and secretive society, part scientific school, part religious community, devoted to the study of numbers, which he considered to be "the source and the root of all things." Out of reverence, the members of the sect never referred to Pythagoras by name but as "the Master" or "that Man."

Lysis had joined the Pythagorean fellowship and in time had become a *mathematiko*, or "student," the higher of the two classes of disciples. These learned the more elaborate aspects of geometry, astronomy, and other sciences from Pythagoras himself. The *akousmatikoi*, or "auditors," on the other hand, were limited to hearing summarized lectures without detailed explanations given by one of the Master's disciples.

The Pythagorean disciples' daily program began with solitary morning walks to some quiet place, such as a temple or grove. They walked alone because they considered it inappropriate to engage in conversation or mingle in a crowd before having attained inner serenity.

After their morning walk they met in groups, in a temple or some other peaceful place, to discuss the Master's teachings or listen to lectures. Next, they turned their attention to the health of the body. Most of them raced, while others wrestled or exercised by jumping with weights.

They lunched on honey and honeycomb and bread made of millet or barley. In the afternoon they resumed their walks in parties of two or three, reviewing the teachings they had received and the precepts they had learned, after which they visited the bath house to cleanse themselves for the evening rituals.

Following their ablutions they gathered in the common dining room, where Pythagoras performed libations and fumigations in honor of the deceased and the gods. They represented the gods not by anthropomorphic images but as the sphere, the divine receptacle in form and nature similar to the Universe. These rituals he repeated three times, because the tripod was the symbol of Apollo's prophetic powers.

For supper they ate raw or boiled vegetables and herbs, cheese cakes, grapes and dry figs, very rarely fish and never beans or animal meat, and they drank wine. The evening ended with the reading aloud of some of Pythagoras' precepts and maxims, such as: *Do not devour your heart* (meaning, we ought not to afflict ourselves with grief and sorrow). *Do not poke the fire with a sword* (Do not provoke a man in anger). *Do not negligently enter into a temple, nor adore carelessly. Always sacrifice and adore barefoot. Make thy libations to the Gods by the ear* (Beautify thy worship with music). *To the celestial Gods sacrifice an odd number, but to the infernal, an even* (To God consecrate the indivisible soul; offer the body to hell). *Leave not the least mark of the pot on the ashes* (After reconciliation, forget the disagreement). *Help a man to take up a burden, but not to lay it down* (Encourage not idleness but virtue). *Write not in the snow* (Do not trust your precepts to persons of an inconstant character). After which they separated and went home.

As Lysis knew well, certain Crotonians were hostile to the fraternity and capable of base and treacherous acts, just like Hipparchus, a former disciple of the Master, who had philosophized indiscriminately and publicly when the very first thing the Master taught was that his doctrines should be preserved in silence and under no circumstances revealed to the profane. He had since been expelled from the fraternity and a symbolic tomb had been built for him with "Let him be declared dead" written on it.

Still worse had been the fate of another sacrilegious disciple, Hippasus, who had divulged the secret of the incommensurable quantities to those unworthy to receive it. So indignant with him was the Divine Power that it unleashed a terrible storm while he was sailing at sea, and the impious perished when his ship was wrecked on the sharp reefs of the Ionian coast.

Two line segments are said to be *incommensurable* (i.e., without common measure) if the ratio of their lengths cannot be expressed as a fraction n/m of integers n, m. The Pythagoreans were the first to discover that the diagonal of a square is incommensurable with its side. Put another way, if the side of the square measures, say, n centimeters, then the diagonal is not an exact number of centimeters; and the same is true regardless of the unit of length used. In practice, a "common measure" of any two segments can always be found, within the limits of precision, by choosing the unit of length sufficiently small. This empirical fact may have led to the belief that segments were always commensurable.

How the Pythagoreans, or rather some unknown Pythagorean, discovered the existence of incommensurable lengths has been the subject of much speculation. But a passage in Aristotle suggests a possible argument leading to the result: Aristotle says that the diagonal cannot be commensurable with the side because if it were, then the notions of odd and even numbers would coincide. The argument hinges on the early Pythagoreans' conception of a line segment as made up of a finite number of points and the fact that their geometry was based on the natural numbers.

Suppose then that we construct a square whose side has an odd number k of points. How many points does its diagonal have? Say that the number of its points is n. The unknown Pythagorean must have deduced that n should be even, because according to his Master's famous theorem, $n^2 = k^2 + k^2$, so the square of n equals $2k^2$, which is even, and only even numbers have even squares. But, on the other hand, n would have to be odd, because if n is even, its square must be a multiple of 4, which $2k^2$ isn't.

At this point the Pythagorean would have been at a loss to make sense of a number that was at the same time even and odd—a logical impossibility. (For the Pythagoreans, whether a number was even or odd was important, as this property was related to other, nonmathematical ideas.) And so, our unknown Pythagorean—we imagine him young and blond, with a sharp intellect and a predisposition for inquisitive speculation—would have concluded that the diagonal *had no number!* It could not have occurred to him that the way out of the paradox was to realize that the length of the diagonal was a *new kind* of number we now call *irrational* (i.e., not expressible as a ratio n/m of two integers). We write this number as $\sqrt{2}\ k$, where $\sqrt{2}$ (the square root of 2) is the (irrational) measure of the diagonal using the side of the square as unit of length.

From the vantage point of our present-day knowledge, we can hardly imagine the anguish and despair the discovery of incommensurable quantities might have caused the Pythagoreans. Apart from the fact that it dealt a blow to their tenet that "all is number and proportion," inasmuch as mathematics was for them a description of reality, they may have glimpsed in this discovery a contradiction at the heart of reality itself. So terrible a truth had to be hidden at all costs. What if they had stumbled upon some forbidden knowledge the gods never intended humans to posses? Who knows what unspeakable punishment might befall them? Wasn't Hippasus' death at sea a retribution from the divinities for having revealed the secret?

On the other hand, the mind of the unknown Pythagorean must have been blessed by the gods themselves, for he had not merely discovered a new truth, he had done it by the power of reasoning alone in what is probably one of the first examples of a mathematical proof. The mathematics of his predecessors in Babylonia or ancient India is rich in arithmetic and geometric facts; but these are only stated, never proved by means of a logical argument. It is the presence of proofs from first principles or axioms that sets Greek mathematics apart, in particular those in Euclid's *Elements*, which became the standard for mathematical rigor for nearly two millennia.

Among the most prominent citizens of Croton was a wealthy aristocrat by the name of Cylon, a man with a disposition to take offense easily and to react violently, and who would not stop at anything to achieve his ends. As he considered himself worthy of whatever was best, he deemed it his right to be admitted to the Pythagorean fellowship, and he therefore went to see Pythagoras for that purpose.

Those wishing to enter the fellowship had to pass a certain initiation test, which could take up to one year to complete. But prior to the admission test Pythagoras would study their manners, their gait, and the motions of their whole body, for he considered these as visible signs of the invisible tendencies of the soul, such as the potential to engage in serious studies or the ability to cope with the rigors of an ascetic life.

It did not take very long for Pythagoras to discern Cylon's true nature, and to tell him in no unclear terms that his admission to the fellowship was out of the question. The rejected applicant took this as a great affront and became furious, and, just as Lysis feared, devised a vindictive scheme to get back at Pythagoras and his disciples.

His opportunity came following the defeat and capture by Croton of its rich and luxurious neighbor, the city of Sybaris, on the Gulf of Tarentum, after a ferocious battle. For many years the Crotonians had enjoyed peace and prosperity. But the people had changed, and they were no longer content with the old magistrates and the original form of government. After the capture of Sybaris, the dissatisfaction of many with the way the conquered land had been divided provided a pretext to challenge the existing constitution. Lysis had been present at the assembly where changes to the constitution had been debated, such as opening the magistracy to every citizen. When this was opposed by the Pythagorean Democenes, Cylon took the opportunity to mount an attack on the fraternity. In a long and fiery speech he accused the Pythagorean philosophy of being a conspiracy against democracy.

After Cylon's defamatory tirade had aroused the masses, the plebeian Ninon had followed with his own calumnies, pretending he had penetrated the secrets of the Pythagoreans. A scribe had then begun to read from a book given to him by Ninon entitled *Sacred Discourse*: "Friends are to be venerated in the same manner as gods, but others are to be treated as brutes." A series of other lies and fabrications had followed: that Pythagoras had praised Homer for calling a king a shepherd of the people—an implicit approval of aristocracy in which rulers are few, while the implication is that the rest of men are like cattle; that beans should be scorned because they are used in voting (voters signified their approval or disapproval of a candidate for public office by placing a light-colored or dark-colored bean in an urn, respectively), whereas the Pythagoreans selected officeholders by appointment; that to rule should be an object of desire, because it is better to be a bull for a day than an ox for life; and that the people should remember that when they raised their right hand to vote, that same hand was rejected by the Pythagoreans, who were aristocrats.

The purpose of the joint attacks by Cylon and Ninon had been clear: to stir up the Crotonian people against the Master and his disciples. The worst is still to come, thought Lysis; there is no telling to what extremes Cylon might go to seek revenge, and now he has the masses on his side. He feared that a revolt against the Pythagoreans was in the making. It was no doubt on this account that the Master wished to see him.

When Lysis arrived at the house he was greeted on the porch by Myia, Pythagoras' eldest daughter. "The Master awaits you," she said, as she held open the large door leading into a high-walled inner courtyard. As Lysis walked across it he passed the altar dedicated to Zeus Herkeios, "the protector of the walled courtyard," on his way to the *andron*, or reception room.

Zalmoxis was standing in the small antechamber. Without speaking a word, he pointed to the adjacent room, whose only light came from the courtyard through a small window high on the wall. Lysis stood in the doorway, waiting for a signal to step inside. His eyes, slowly adjusting to the darkness, could barely make out the long-haired figure of the old man in the immaculate white *chiton*, an ankle-length tunic, sitting on a wooden bench at the opposite end of the room.

"Come in and be seated, my loyal friend," said Pythagoras at last. He had been studying Lysis' expression. "I see troubling thoughts crossing your mind."

"You see right, Master," replied Lysis.

He sat down on a low stool before speaking again.

"Cylon is in a vindictive mood. He has aroused the assembly against the fraternity and violence may result."

"To act inconsiderately is part of a fool," was Pythagoras' calm response.

But Lysis had still another source of concern: "I have been having some disturbing dreams of late, premonitory dreams, I believe. I fear for your life, Master."

"Don't forget that death is appointed to all . . ." And then he added, in a tone that had suddenly become mellow: "I have had some premonitions of my own, and they are the reason I have sent for you, Lysis, my most faithful and dependable friend."

Lysis' only reaction was a slight bow of the head. He was flattered by the compliment but would not let his expression show it.

"What I shall ask you to do is known to my daughter Myia, but to no one else, not even to my wife Theano or any of my most intimate friends. No other person besides us three shall be let in on the secret; no one, that is, until the right man comes along."

He stroked his white beard in a pensive manner, increasing the dramatic effect his words were having on Lysis.

"The Providence has spared me so far," Pythagoras resumed, "but I have now received some signs from the gods that my soul will soon journey away from my tired body and migrate to another one in order to accomplish a mission of the utmost importance, and I need your help to carry it out."

As he spoke, he rested his left hand on an object that had been lying on the bench beside him but that only now Lysis remarked: a metal cylinder, of the kind used to keep papyrus scrolls.

"It's all written here, in this scroll," said the old man, tapping on the metal case, "and I must absolutely avail myself of it on my return among the living." There was a short pause before he added: "You are to follow my instructions to the letter, lest this precious document should be lost or, even worse, fall into the wrong hands."

Pythagoras then instructed Lysis to keep the scroll in a safe place and guard it jealously, and to make arrangements so that, upon his death, its safekeeping be transferred to his son, or daughter, or wife, or a trusted friend. This person would have to promise allegiance by taking the Pythagorean Oath: "I swear by he who revealed to our soul the Tetraktys, that most sacred symbol, the source of all our wisdom and the perennial root of Nature's fount." The new keeper was to pass the document on to his own descendants, who would hand it down to theirs and so on from generation to generation.

The successive keepers would enjoy the protection of Apollo, but they should be warned that the wrath and the curse of the god would fall upon them should they neglect their duties as custodians of the precious manuscript, break its protective seal, or otherwise put in jeopardy the success of Pythagoras' mission. The passing of the scroll from one keeper to the next should continue, perhaps for a long time, until the reincarnation of Pythagoras was ready to carry out his mission. The Providence would then send a signal to the current keeper to deliver the scroll to him.

Lysis had been listening attentively to the Master. He was proud to have been chosen as custodian of the scroll but realized the heavy responsibility that had befallen him. Although he wished the Master would tell him a bit more about the nature of the mission, he did not dare to ask. But he felt that one particular point required further clarification: "When the appropriate time comes, how, among all men, will the custodian of the scroll recognize the Master?"

Pythagoras seemed annoyed by the question, as if his disciple should have known the obvious answer. But there was no sign of irritation in his voice when he replied, fixing Lysis as he spoke: "He will be an extraordinarily gifted man, eminently versed in the secrets of Number, of whom many wonderful things will be persistently related."

Chapter 9

NORTON THORP

N orton Thorp was born in Moscow in the spring of 1963, the only child of a middle-aged American diplomat who was serving his country in the Soviet Union during the cold war and a much younger Russian ballerina. Norton's parents never married, and when Donald Thorp left Moscow to return to the United States, taking his four-year-old son with him, Marina Golikova, the child's mother, stayed behind, ostensibly for professional reasons. And so Donald's younger sister Therese, a divorcee with no children of her own, became for all practical purposes Norton's second mother—and part-time father as well, for Thorp's assignments abroad kept him separated from his son most of the time.

Therese was comfortably well off thanks to the generous divorce settlement that her attorney and for a while also lover, Morris Pringley, had managed to extract from her former husband. She was a pleasant, cultivated woman in her thirties with no particular talents or ambition but with a cheerful disposition that the breakdown of her marriage had not dampened. In her providential new role as Norton's surrogate mother she appeared to have found her purpose in life, and henceforth devoted herself to her nephew's upbringing with the zeal and abnegation of a missionary.

Norton was no ordinary child. From the moment he was reunited with his mother—his premature birth had confined him to an incubator for three weeks—he began to show unmistakable signs of precocity. He was an extremely alert baby, who slept little and cried even less, his sparkling green eyes constantly scrutinizing his surroundings, absorbing every detail and keeping track of the slightest movement, and with his attention always sharply focused as soon as anyone talked to him. At nine months, he talked in complete sentences. He would speak in Russian with his mother and his live-in nanny but would only use English to communicate with his father. Donald would often read aloud to his son, who would sit up quietly beside him and insist on looking at the written words as if to check the accuracy of his father's version of the story.

By the age of three Norton spoke English and Russian fluently, with a vocabulary well beyond what might be expected from a child or even a young adult, and by the time he started kindergarten he could already read English without hesitation. He kept in touch with his biological mother mainly by telephone, and the monthly conversations the two of them had carried on separated by an ocean and a continent helped to refresh his Russian. He also acquired a good knowledge of French from some old picture books that had belonged to his paternal grandmother Thérèse-Marie Thorp (née de Sèvres), a French aristocrat after whom his aunt had been named. Soon after majoring in French literature at the University of Michigan, the young Therese had spent one year in Paris perfecting her French, and she now welcomed the opportunity to have someone, even a young beginner, with whom she could speak it once again.

Music, especially but not exclusively classical music, was a central element of Therese's universe. She would invariably have the radio or the four-track stereo system playing as she went about her daily chores and also in the evenings, which she usually spent alone reading or watching television while the music played in the background. She was vice president of the Ann Arbor Baroque Music Association and regularly attended concerts and other musical events but, apart from a few appearances with the university choir in her student days, she did not feel the need to experience music as a performer. Despite having been brought up in such a propitious environment, Norton never manifested any real interest in music, let alone in playing an instrument. Some gentle nudging on the part of his aunt, who would have liked him to take piano lessons, produced no results.

Morris, Therese's former lover, would occasionally drop in and more often than not also stay for dinner, but he would no longer stay overnight. Their brief affair, now a thing of the past, had been replaced by a passionless but sincere companionship in a mutual attempt to fend off an all-too-real sense of loneliness.

Therese was an excellent cook. Her French background probably had something to do with the ease and assurance with which she could combine the most heterogeneous ingredients and produce a surprisingly delicious dish that she was happy to share with a delighted Morris. The young attorney was no gourmet but he could certainly appreciate a fine meal, especially one that couldn't compare with the typical American fare. It was after one of those intimate dinners that the first of the "incidents," as they would later refer to them, occurred.

Morris had just turned thirty-five, and to mark the occasion Therese had decided to cook him something special: a four-course Middle Eastern meal. The afternoon rain had cooled the air and the late autumn evening threatened to be rather chilly but the table for two had nevertheless been set outside, on the patio adjacent to the large kitchen where Therese had once again worked her culinary magic.

Young Norton—he was then barely five—had already been fed, bathed, kissed good night, and tucked up in his bed for the night when Therese and Morris sat down to enjoy their exotic feast. It opened with the appetizers, an assortment of dips that included grilled eggplant and lemon puree, a spread made from feta cheese spiced with chili pepper and garlic, and meat cooked in tomato and red wine sauce. An entrée of burghul and potato cakes with lamb and apricot filling was followed by the main course: swordfish baked in a lemon and paprika sauce and served on a bed of pilaf rice.

Wine had been steadily flowing throughout the extravagant meal. Dessert—figs in their syrup topped with whipped cream and pistachios—was still to come and the mood was light around the candle-lit table.

"So where did you dig up the recipes for those exotic and probably aphrodisiac dishes? And don't tell me you made them up yourself." Morris was teasing her, as he usually did when he was slightly intoxicated.

"No, I didn't make them up—this time. And I didn't add any funny stuff either. I found the recipes in the book of the One Thousand and One Nights" she replied, quite happy to play the game.

"C'mon, there are no recipes in the Arabian Nights book, only the stories that Scheherezade spins so that the king's curiosity will prevent him from having her head chopped off."

"Oh, yes there are—in the original Persian version, that is. Scheherezade not only tells King Shahryar fabulous tales; she also cooks the most exquisite dishes to keep him happy. In fact, she owes her salvation to her cooking rather than to her storytelling. It's there, recipes and all, in the Persian version. But it got lost in the modern translations."

"And you would want me to believe that you can read Persian?"

Under the starry sky, the patio was now almost completely dark. The kitchen lights behind them had been dimmed to the point of extinction and only the wavering flame of the half-consumed candle on the table illuminated the playful exchange of pleasantries between the two friends. They could hear the music, a piano sonata or something of the sort, coming from the patio of the adjoining semi-detached cottage.

Therese was the first to realize that something was not right. John and Ethel, the elderly couple living next door, were out of town for the week—she had gone over in the afternoon to water the plants and check on the house. There were no neighbors around; the music came from inside the house—her own house. She remembered having played a record of Arabic music to add to the atmosphere, but the portable record player had automatically shut itself off, and besides—and this was the most disturbing part—what her well-trained musical ear was detecting was not a recording but the live sound of a real piano.

When Morris finally caught up with Therese she was standing in the living room partially blocking his view, but in the diffused lighting he could still make out the silhouette at the back of the room. In a corner, near the bay window, there stood a boudoir grand piano that Therese had inherited from her mother. The massive black instrument had been silent for years but not totally neglected, for Therese had it dusted every fortnight and occasionally tuned, out of a self-imposed obligation to properly care for the instrument that had meant so much to her mother. And now *someone*—a very small someone—was playing it.

"Norton? What . . ." Therese could not finish the sentence. She was trying to decide which one of her senses not to believe: her hearing or her sight. Norton was sitting on the piano stool in his bright-colored pajamas, his little legs, too short to reach the pedals, dangling in the air while his small hands flew over the keyboard, pressing and releasing keys with the skill and assurance of an accomplished pianist. Therese and Morris just stood there, bewildered and petrified, listening to Norton's consummate rendition of what some part of Therese's brain recognized as the third movement of Mozart's piano sonata K.331 in A major, popularly known as the "Turkish Rondo," with its typical broken chords lending emphasis to the main beats.

When they later talked about the incident, they could not agree on how long it had lasted but they did remember having been mere witnesses, two silent, motionless, and unbelieving witnesses, throughout the whole unreal experience. They also recalled how it had ended: Norton climbing down from the stool after he finished playing the piece, holding on to the seat while his bare little feet reached the carpet, and then picking up the toy giraffe he had abandoned near the piano, curling up on the carpeted floor, and finally falling asleep hugging his stuffed companion.

After much reflection, hesitation, and discussion with Morris, Therese decided against telling Donald, at least until she could figure out some kind of explanation for what had taken place that night. But the true reason for her silence was a fear of being separated from Norton, since her brother might not wish to continue to entrust his son to someone prone to nocturnal hallucinations, caused by god-knows-what intoxicating substance.

The day following the incident Norton was his usual good-humored self, eating, playing, and generally behaving in the most normal fashion. When Therese played a recording of Mozart's Turkish Rondo on the stereo, making sure that Norton was around to listen, the boy's reaction was no different from his customary response to music in general: one of total indifference.

The much sought-after explanation would never be found, their combined efforts to shed some light on the incident notwithstanding.

Therese read a great deal about visions, hallucinations, and paranormal phenomena and the biography of Mozart and other infant prodigies; she even delved into pseudo-scientific and esoteric literature on spells and witchcraft, but stopped short of asking her parish priest about the possibility of Norton having been possessed by some evil, if talented, spirit.

Morris, for his part, conducted his own inquiry into the intricacies and unpredictability of the human brain, looking for a physiological or psychological explanation for the mystery. Without disclosing too much about what had really happened, he had tried, unsuccessfully, to have a psychiatrist friend of his prescribe an encephalograph for Norton, hoping it would reveal some physical clue. He had also suggested questioning the boy under hypnosis but Therese would have nothing to do with it, fearing the adverse side effects it might provoke in the child. She had always considered hypnosis a kind of mental rape that could permanently alter the subject's personality.

Two years after the incident Therese and Morris had gradually resigned themselves to remaining forever ignorant of its true nature. The unexplained event had become their bonding secret, and with the detachment afforded by time they were now able to joke about it. "Maybe it was our conversation about the one thousand and one nights. We may have accidentally conjured up some genie who granted your wish that Norton played the piano," said Morris. "At least you are no longer blaming my cooking," replied Therese, "even if you have decided to stay away from Middle Eastern food." They laughed. Given that nothing out of the ordinary regarding Norton had happened since that special night, they had reached a kind of closure—a feeling that was to be shattered some ten years later, when the second "incident" occurred.

Chapter 10
RANDOM NUMBERS

Johanna Davidson's fascination with randomness dated back to her first course in probability and statistics. What she had found most intriguing was the fact that the teacher could not provide a satisfactory definition of "random" (or of "probability," for that matter), even though notions such as "random variable" and "random sample" lie at the heart of the theory. She would later learn that this was not due to the lecturer's imperfect knowledge of the subject, but rather to the difficulty of specifying what, for example, a random sequence of 0s and 1s precisely is. In fact, this question has a long history and no definitive answer.

A simple—if not very practical—way of generating a string of 0s and 1s in a random fashion is to repeatedly flip a perfectly balanced (or "fair") coin and to write down the outcomes as "1" for heads, and "0" for tails. Thus, after ten tosses of the coin, we may have obtained the sequence

$$0 1 0 0 1 1 0 0 1 0$$

whose binary digits (or *bits*) are totally unpredictable: the eleventh bit is equally likely to be 0 or 1, and the same is true for any subsequent bit. On the other hand, sequences that exhibit some kind of pattern or regularity, such as

$$0 1 0 1 0 1 0 1 0 1 \ldots \text{ or } \quad 0 1 1 0 1 1 1 0 0 1 0 1 \ldots$$

would not be considered random, for we can predict what the next bit will be. In the first sequence on page 69, 0s and 1s alternate, so the eleventh bit will certainly be 0, the twelfth will be 1, and so on. The other sequence is constructed by writing the numbers 0, 1, 2, 3, 4, 5, ... in base 2, one after the other, that is, 0, 1, 10, 11, 100, 101, etc., so that we can tell what bit must appear in position $n + 1$ by looking at the first n bits.

Two fundamental questions regarding binary sequences are (1) deciding whether a given sequence of 0s and 1s is random or not, and (2) determining how to generate random sequences. Answers to these questions have become increasingly important with the pervasive use of computers to simulate complex phenomena, such as the interactions of subatomic particles, the evolution of galaxies, and the consequences of the malfunction of a nuclear reactor. Such simulations may require millions of random numbers a minute, and having a reliable source of these is crucial for the validity of the final results.

In 1992, Alan Ferrenberg and David Landau, physicists at the University of Georgia, and Joanna Wong from IBM were using a computer simulation to study the property of certain materials to suddenly enter into a magnetic state when cooled below some critical temperature. The atoms in a single layer are simulated by points on a grid, whose color represents the atom's magnetic moment orientation (north or south). The simulation chooses atoms at random and decides whether they should remain in the same magnetic state or switch to the opposite state. The probability of a particular atom changing state depends on the state of its neighbors and the temperature. The computer program calculates this probability and then compares it with a random number between 0 and 1. If the random number is larger than the probability, the atom remains in its current state; otherwise it flips to the opposite state.

The program seemed to be working smoothly, except for one problem: the computer model consistently predicted the wrong temperature for the transition of the material to a magnetic state. The researchers finally tracked down the culprit: the program's new random number generator. This generator was faster and used more advanced mathematics than conventional ones, and it had passed tests indicating that it produced sequences of numbers that were closer to genuine randomness than those obtained from standard generators. But, as they discovered, the new generator behaved worse in practice.

Johanna had read a report of the incident in *Physical Review Letters*. It was a confirmation of her own concerns about the possibility of hidden flaws in the random number generators on which most simulation programs rely. Her line of work as a systems analyst was not directly related to random numbers but dealt instead with network security—how to prevent intruders from breaking in and stealing or corrupting data stored in the network computers or otherwise disrupting their normal operation. But since randomly generated encoding keys are the most secure way to protect encrypted information, there might be a connection after all.

She had gotten in touch with Ferrenberg, who had told her that other scientists had contacted him with worries about their own simulations. "This is a warning bell," Ferrenberg had said. "Unfortunately, it appears that ostensibly high-quality random number generators may lead to subtle but dramatic errors in some algorithms."

Ferrenberg and his colleagues' work in the physics of solids was mainly theoretical. But computer simulations were confidently and routinely applied to a variety of fields and situations, resulting in decisions that had practical implications well beyond the laboratory. Computer models were used to predict the consequences of global warming, to anticipate the force and the path of tornadoes, and to study the progression of epidemics; advanced computer programs could now simulate the detonation of a nuclear device on a supercomputer—a much more economical and safer alternative to an actual nuclear test.

Most of these computer simulations rely at some stage or another on the automatic generation of random numbers in order to incorporate the elements of chance and uncertainty present in the very fabric of physical phenomena. When billions of those random numbers are employed to simulate the evolution of a complex system, the accuracy of the final results may directly depend on the quality of the "randomness" built into the program. To make matters worse, as the incident reported by Ferrenberg illustrates, random number generators which seem to work well in test runs may turn out to be unreliable in practice or when used over long periods of time, and there may be no way of knowing in advance that the results of the simulation might be flawed. But what is a true random sequence, and how can we manufacture one?

Several definitions of "random sequence" were proposed in the first half of the twentieth century,* but none of these captured the essence of the concept. Finally, in 1965, Per Martin-Löf, a young Swedish mathematician, came up with a sophisticated definition that appears to be the right one. His definition can be better understood through its connection with a concept developed by the Russian mathematician Andrej Kolmogorov. In the 1970s, Kolmogorov defined the "complexity" of a finite binary sequence as the length of the shortest computer program that can print it. For example, a sequence of one million 1s has very small complexity, for there are very short programs that can be used to print it—for instance, in many computer languages, something of the form: "for $i = 1$ to 1,000,000, print '1.'" Such a program may be considered a "compressed" version of the given sequence. On the other hand, a sequence of 1,000,000 bits whose Kolmogorov complexity is 1,000,000 is totally "incompressible": there is no shorter way of describing it than listing all its one million bits.

Now, "randomness" and "incompressibility" turned out to be equivalent concepts, that is: an infinite binary sequence is random (in Martin-Löf's sense) precisely if it is incompressible. An immediate practical consequence of this result is that no computer program can generate a truly random binary sequence. The reason is simple: any sequence s whose bits are obtained by executing a computer program is automatically compressible (for the program is a compressed version of s) and hence it is nonrandom. It follows that all random number algorithms being used in computer programs are in fact only *pseudo* random, and not simply for lack of ingenuity on the part of the mathematicians who created them but due to the existence of an essential barrier inherent in the concept of "randomness" itself.

Johanna was aware of this theoretical obstacle, but had always felt that there might be a way to circumvent it. The situation was made even more puzzling by the fact that, even if "almost all" binary sequences are random, no algorithm or computer program could produce a

*See appendix 3.

single one of them. Randomness is everywhere and yet it is out of reach!

There was another aspect of reality, as Johanna knew well, where randomness was pervasive: the realm of quantum mechanics. In classical physics, determinism and causality reign, while probability is only a convenient way to make up for our incomplete knowledge of the acting causes. Even the outcome of a coin toss is in principle predictable: if we knew all the variables involved and their initial conditions, the final position of the coin would be completely determined by the equations of motion. In the quantum world, on the other hand, events occur at random, and probability is no longer a substitute for our imperfect knowledge of the underlying causes. There is no "cause" behind the decay of an atom from an excited state, only chance; and the "laws" governing the process merely express the probability of the event taking place at some particular moment. This was Max Born's idea in the 1920s, fully confirmed by innumerable experiments since, but rejected at the time by Einstein. In a letter to Born he wrote that quantum mechanics was very imposing, but he was convinced that God did not throw dice.

On a bitterly cold morning in February 1998, Johanna Davidson woke up with a mild headache. A few minutes passed before she reluctantly crawled out of bed, pulled up the blinds, and looked out the window. Snow had been falling steadily over Boston since the day before, leaving the city half buried under a soft white mantle. In the street below, traffic was barely moving, while pedestrians struggled to walk in the fifteen-inch-deep snow that had accumulated on the sidewalks.

Johanna lived alone in a one-bedroom apartment on the second floor of a century-old brownstone building located in the fashionable Back Bay district of the city. A spacious kitchen with exposed brick walls, a fully equipped bathroom, and an oversized living room made up the rest of the unit.

Her morning headache was probably the result of a bit of alcohol overindulging the night before, but it could equally well have been due to the stormy and abrupt end of her evening out with Kevin—and of her having to face the prospect of yet another relationship hitting the rocks.

At thirty-four years old, Johanna was still in her prime. She had a graceful figure of medium height, curly brown hair, and a bright, intelligent face with a delicate nose and engaging gray eyes. Professionally, she was a successful computer consultant, traveled extensively on business, and had no financial worries, but in her sentimental life the "downs" largely outnumbered the "ups." Not that she was really looking to get married and start a family. She wasn't *necessarily* looking for that, but she would have preferred it to be her decision, instead of the choice being forced on her by default, for lack of a suitable partner.

She showered, fixed herself a bowl of cereal with plenty of milk, and took it with her into the living room. A good half of the spacious room was arranged as an office. The furniture included an old oak desk, two file cabinets, and a huge bookcase overflowing with books. On a large, sturdy table along one wall were two computers with widescreen monitors, a laptop, a printer, and various other peripherals.

She sat down at one of the computers and logged on to a bulletin board on virus alert she had been running; she wanted to see if it was still up. Hackers had managed to shut it down a couple of times but this morning the site was working smoothly; the extra protection she had added seemed to be keeping intruders at bay. Then she checked her e-mail. There were five new messages—none of them from Kevin. Her face lit up when she saw the message from Andrew Stone. Andy, as he urged everyone to call him, was her former PhD thesis supervisor and lifelong mentor. Now a retired professor of computer science from McGill University, in Montreal, he was still as active as ever. Released from teaching and administrative duties, his seemingly inexhaustible energy now manifested itself through other channels.

She clicked the message open. *Hi, Jo. Where are you this time? I thought you'd be interested to know that Norton Thorp has accepted my invitation to give a talk at our joint Math-Computer Sci. seminar. Hope to see you there. Love, Andy.* Details of time and place followed.

Of course she was interested. Norton Thorp needed no introduction, and Johanna wondered how Andy had managed, even taking into account his considerable persuasive powers, to entice such a scientific superstar, a mathematician in a class of his own, to give a talk at McGill. To be sure, the university had an excellent reputation, and its department of mathematics probably ranked among the top twenty in North America. But Thorp enjoyed such an international celebrity standing that he was sought after by universities, research institutes, and scientific laboratories all over the world. Invitations to give a lecture, chair a conference, or deliver an opening address—not to mention requests for interviews and television appearances—would be bombarding him as steadily as spam e-mail. And yet, against all reasonable expectations, Thorp had decided to accept Andy's invitation to come to Montreal.

Johanna was not about to miss the opportunity to hear him lecture and perhaps learn a thing or two about random number generation—a field, like so many others, to which Thorp had made significant contributions.

Her thoughts drifted toward her twin brother Jule, who was also a mathematician. She had been thinking of him often lately. Not that she was really worried, but Jule had been behaving in a strange way, beginning with his taking leave from the university for what he called "a special assignment" in Chicago. At first he was evasive as to the nature of his assignment, but when finally, at her insistence, he told her all about it, she couldn't believe it. She thought the whole thing was crazy.

Chapter 11
RANDOMNESS EVERYWHERE

McGill University's main campus is located in downtown Montreal, at the foot of the Mont Royal hill, the city's principal landmark. Just beyond the semicircular stone-and-iron entrance gate and on the right sits Burnside Hall, a box-shaped concrete building housing the mathematics and statistics department.

At two o'clock on a sunny afternoon in mid-April 1998, Burnside Hall's main auditorium was almost packed, even though the lecture was not scheduled to start until 3:30. The speaker's reputation had attracted an unusually large crowd for a mathematics talk. Its title, "Randomness at the Heart of Mathematics," had aroused the interest of local and out-of-town mathematicians and computer scientists, who suspected that Norton Thorp would announce a breakthrough in the generation of random numbers, a subject central to the computer simulation of real-world phenomena.

By the time Andy Stone stepped on stage to introduce a speaker who needed no introduction, every available seat and standing space in the auditorium was occupied. Johanna Davidson was sitting in one of the front rows. She had arrived early, hoping that Andy would introduce her to his famous guest and she could have him sign her copy of *Life of a Genius*, his recently published biography. But her former teacher was too busy looking after the VIPs invited to the talk, and she obviously was not among them.

Norton Thorp was an artful speaker, and he knew how to structure his lectures in order to maintain the interest of his audience. He would begin with some amusing anecdote or clever story to break the ice and then gradually step up the level of technical difficulty until he reached the climax, some deep new formula or theorem that only a few would understand. But he would then quickly climb down from those conceptual heights to wrap up the talk with some general remarks in plain language about the significance of the new result, often accompanied by philosophical considerations that everybody could appreciate.

As he appeared on stage and approached the lectern, Thorp was received with a sustained round of applause. He thanked the organizers and especially Andrew Stone for the invitation, said a few words about how delighted he was to be in Montreal, a city with "a blend of French and English cultures unique in North America," and began his talk.

"Computers are supremely fast for doing arithmetic, and they can easily beat humans at the task of, say, adding five hundred numbers. But how about adding infinitely many numbers? Let me tell you a little story.

"Suppose we wished to find the result of alternatively adding and subtracting all the reciprocals of the odd numbers, an infinitely long operation

$$1/1 - 1/3 + 1/5 - 1/7 + 1/9 - 1/11 + 1/13 - 1/15 + \ldots$$

The numbers appeared on the overhead screen.

"We could use a computer to do the calculations, but one thousand years and zillions of operations later, the machine will still be trying to work out the final answer—although of course, none of us will be around by then." There were some laughs from the audience. "OK, let's be fair to the machine and admit that at that point it will have found an answer," Thorp resumed, "but only a partial and approximate one."

"In 1671, centuries before the first computers were built, the Scottish mathematician James Gregory proved that the answer to the infinite operation is *exactly* $\pi/4$, establishing at the same time a beautiful and intriguing connection between elementary arithmetic—whole numbers, addition, subtraction—and elementary geometry—the ratio of the circumference of a circle to its diameter. The moral of the story?

A machine can add billions of numbers, but it takes a human brain to calculate an infinite sum."

He struck a key on his laptop and "Randomness at the Heart of Mathematics" flashed on the screen. The talk had changed gears.

"The theory of probability is the best tool humans have devised to deal with random phenomena, that is, those situations involving chance. The simplest example of a random event is the result of flipping a 'balanced' or 'fair' coin, which can land either 'heads' or 'tails.' If we repeatedly toss the coin and record each successive outcome by writing 1 (for heads) and 0 (for tails) we'll end up with a sequence of 0s and 1s, such as, for example, 0 0 0 1 1 0 1 1 0 1. All sequences obtained in this way qualify as 'random sequences.' This means that chance alone decides the composition of the sequence—and not, say, some predetermined rule."

After this elementary introduction he recalled the various attempts made by mathematicians to specify what a random sequence of 0s and 1s precisely is, and he reviewed the most efficient algorithms used to generate pseudo-random sequences on a computer. "A true random sequence is totally unpredictable," he went on, "it doesn't obey any law; it is not just that we are not smart enough to figure out the rule or pattern; *there is no rule*, and in this sense true randomness is indistinguishable from absolute chaos. It is precisely because of this complete absence of any order or structure that a true random sequence cannot be described other than by writing down the entire sequence.

"But how pervasive is randomness? To answer this question, we first need to make a detour through the theory of computation." The pleasant-looking face of a young man deep in thought appeared on the screen.

"In the summer of 1935, a young Cambridge graduate reflecting on a question in the foundations of mathematics came up with the notion of an ideal computer that could mimic the operations of any computing device. The young graduate was Alan Turing—the brilliant British mathematician behind the cracking of the German Enigma code during the war and considered the father of the modern computer. His ideal computing machine became known as a Turing machine.

"Since Turing's ideal machine is at least as powerful as any real one, anything a Turing machine cannot do, no real computer—present or future—will be able to do either. As Turing showed, one of the problems his ideal computer cannot solve concerns the automatic checking of computer software. This is the question of determining in advance whether any given computer program, when executed, will eventually terminate its calculation and halt or is destined to run forever—the so-called halting problem.

"Many years later, starting in the 1970s, Gregory Chaitin, a mathematician working at IBM, took a fresh interest in the halting problem. He considered all the programs that could be run on a Turing machine and asked the following question: What is the probability that one of these programs chosen at random will halt? He found that the answer is a number, which he called Omega, whose digits form a true random sequence, in the sense that they have no pattern or structure whatsoever. Chaitin described Omega as a string of 0s and 1s in which each digit is as unrelated to its predecessors as one coin toss is from the next. He presented Omega as the outstanding example of something in mathematics that is uncomputable, and therefore unknowable.

"But Chaitin did not stop there. He began to search for other places in mathematics where randomness might crop up, and he found that it does in its most elementary branch: arithmetic. But if there was randomness at the most basic mathematical level, his hunch was that it must be everywhere, that randomness is the true foundation of mathematics. In Chaitin's own words: 'God not only plays dice in physics but also in pure mathematics. Mathematical truth is sometimes nothing more than a perfect coin toss.'

"Was Chaitin right? For many years, the question remained open, but some recent work by two mathematicians at the University of Berkeley appeared to reinforce the idea that randomness—i.e., unpredictability—is as pervasive in pure mathematics as it is in theoretical physics."

He paused and faced the audience. Every pair of eyes in the room was fixed on the speaker. The silence was so complete that one could hear the noise from the overhead projector's cooling fan. Ten long seconds went by before Thorp spoke again in a solemn tone fitting the occasion.

"I've now found an irrefutable proof that, as Chaitin suspected, randomness is at the heart of mathematics," he announced, "and the implications of this fact are far-reaching. It means that we may be able to prove some theorems, answer some questions, but the vast majority of mathematical problems are essentially unsolvable. It means, as one of my colleagues has put it, that a few bits of math may follow from each other, but for most mathematical situations those connections won't exist because mathematics is full of accidental, reasonless truths. And if you can't make connections, you cannot solve or prove things. Solvable problems are like a small island in a vast sea of undecidable propositions."

As he pronounced these last words, the aerial view of an island with "Solvable Problems" written on it appeared on the screen. Slowly, the camera began to move away, revealing more and more water around the island, until the water, with the inscription "Unsolvable Problems" floating in it, completely filled the screen and the island all but disappeared.

For the next thirty minutes, Thorp presented the broad lines of his proof: screen after screen of formulas and equations, a pyrotechnics of high-level mathematics accompanied by the speaker's comments and explanations. It was a brilliant display of the power of the human intellect dealing with abstractions it had itself created in its quest for understanding.

"The complete proof will appear in the *Annals of Mathematics and Computing*," he announced almost apologetically, as if acknowledging the fact that very few in the audience could have understood all the intricacies of the demonstration from his necessarily sketchy presentation. He then added a few remarks about the consequences of his discovery for other sciences, in particular physics, so dependent on mathematics for the formulation of its theories, and concluded on a philosophical note.

"Our old friend Pythagoras thought he had discovered the key to the mysteries of the universe: the world was ruled by numbers in an orderly and immutable way. 'All is number,' he taught, and so by unlocking the secrets of numbers one could understand anything. He may turn in his grave if he learned that the secrets of numbers are for the most part impenetrable and the fabric of reality is made up of chaos and unpredictability."

A few moments of complete silence followed, as if the audience were paying its respects to the optimistic conception that most—if not all—mathematical problems could be solved, and then it exploded into a standing ovation.

When the applause subsided, Andy Stone thanked the speaker for "sharing your landmark discovery with us," and then announced: "Professor Thorp would be happy to take a few questions from the audience."

Johanna stayed in Montreal overnight, at the house of one her friends from her student days, now a full-time mother of three.

During the long drive back to Boston the next morning, Thorp's closing remarks about Pythagoras came to her mind, and only then did she become aware of the coincidence: Jule's "special assignment" also concerned Pythagoras.

Chapter 12
VANISHED

O n Friday May 8, 1998, the news made the front page of every major newspaper:

Famous Mathematician Goes Missing

The celebrated American mathematician Norton Thorp, internationally acclaimed for his ground-breaking work in mathematics and artificial intelligence, is missing. He was last seen yesterday around noon by Mr. Frank Martino, the doorman of his apartment building. Following what appeared to be a heated discussion with two men in front of the building, Mr. Thorp was forced into a car by the two individuals, Mr. Martino said. He gave a description of the car and the men to the police.

Mr. Thorp didn't show up for his scheduled talk at IBM's Thomas J. Watson Research Center in Yorktown Heights, NY, in the afternoon, and as we go to press he has not been back at his Manhattan residence or communicated with his family.

The police are treating the incident as a kidnapping, even though no ransom note or call has been received by Mr. Thorp's family so far.

PART III

A SECT OF NEO-PYTHAGOREANS

Chapter 13

THE MANDATE

Fearing however, lest the name of philosophy should be entirely exterminated from among mankind, and that they should, on this account, incur the indignation of the Gods by suffering so great a gift of theirs to perish, they made a collection of certain commentaries and symbols, gathered the writings of the more ancient Pythagoreans, and of such things as they remembered. These relics each left at his death to his son, or daughter, or wife, with a strict injunction not to divulge them outside the family. This was carried out for some time, and the relics were transmitted in succession to their posterity.

—From *The Life of Pythagoras*, by the Neoplatonic philosopher Iamblichus (c. AD 250–c. 325)

The news from Croton could not have been more devastating. Archippus was dumbfounded. He sat motionless under the olive tree, one of the hundreds that dotted the hill overlooking the splendid harbor and the gulf beyond. All life seemed to have drained from his body, his left hand covering his face and the other still holding the papyrus scroll that his servant had just delivered.

Archippus was a *mathematiko*, a member of Pythagoras' inner circle of disciples. He had come to Tarentum, his birthplace and one of Greater Greece's principal cities, to attend to some family business that had kept him at his mother's house longer than expected. As he now began to realize, that delay might have saved his life.

The scroll was a letter from Lysis, one of his closest friends. It brought "the saddest of news": the Master had perished. With an unsteady hand and hoping that his letter would reach Archippus before his departure back to Croton, Lysis had written to his friend all about the circumstances that had led to Pythagoras' tragic death. Just as Lysis had feared, the vindictive Cylon had sent his followers to carry out a cowardly attack on the fraternity. The Pythagoreans were holding council at the house of Milo, the famous Crotonian athlete and general of the victorious army that defeated Sybaris, when Cylon's henchmen set the building on fire. In order to provide a means for their master to escape, Pythagoras' disciples had thrown themselves into the flames and made a bridge with their bodies. But the mob outside had blocked the gate, and by the time those trapped inside the house finally forced it open, many had died, either burned or suffocated, the Master among them.

The flames, the scorching heat, the screams; Lysis had gone through all that before in his all-too-real premonitory dreams. How many besides him and Dimachus had managed to escape alive? He didn't know. All he knew was that he was still in danger, for he was convinced Cylon wouldn't stop until he wiped out the entire fraternity.

He had hurriedly left the city, taking with him only a few belongings. "I'm on my way to Thebes," he wrote, "carrying a most precious document in the Master's own hand that he entrusted to me. For reasons I cannot disclose, this scroll must be protected at all costs, and for that purpose I shall need your help." The letter ended with a promise to write again soon and an appeal to his friend not to return to Croton but to join him in Thebes instead.

The sun was rapidly sinking behind the hills, its last rays setting the upper branches of the olive trees ablaze when Archippus got up and began walking back to the house. He had devoted the best part of his life to the fraternity and could hardly contemplate a future without it. But even less could he bear the thought of the Master's teachings and marvelous discoveries being extinguished forever.

Lysis had mentioned a document. Perhaps the Master, anticipating his demise, had left behind instructions for the preservation of his doctrines. Archippus made up his mind: he would go to Thebes to join Lysis. The name of philosophy would not be allowed to perish.

Chapter 14

THE BEACON

When Dr. Gregory J. Trench finished his telephone conversation on an unseasonably warm evening in early January 1998, he was wearing a contented smile. The forty-four-year-old physician had been talking to his friend Leonard Richter, who had told him what he wanted to hear: the team was now complete. The team in question had nothing to do with his medical training and everything to do with his religious beliefs. To understand the connection, one must begin with the story of Trench's conversion to an esoteric faith.

Dr. Trench came from a Catholic family of second-generation Irish immigrants. His grandfather, Tobias Trench, arrived in America from Dublin in 1908 and settled in Chicago, where he opened a small printing shop specializing in religious literature. By the time Gregory was born, in 1954, it was his father, Theodore, who was running the business, which now also included a bookstore: "The Beacon." Besides its stock of religious publications, the store, located in the basement of the renovated family house, also carried a large selection of books on related subjects ranging from history and philosophy to astrology and esotericism.

Theodore and his American wife Mary Ann were both Roman Catholic. But unlike the woman, who observed the rites and precepts of her creed to the letter, Theodore was not particularly devout. His practice of religion seemed more a matter of tradition than conviction. Even though the couple's two boys and two girls had been baptized, the father was reluctant to impose the Catholic faith on them.

"I shall not indoctrinate my children," he would say to a disapproving and disappointed Mary Ann. "They are intelligent enough to find their own way. Why would the Lord have given us a brain if we were not supposed to use it?" Far from indoctrinating them, he incited his offspring to be wary of those preaching "the final and absolute truth" in religious or secular matters, and to be open-minded and trust their own judgment.

Gregory was especially receptive to his father's advice. An avid reader at an early age, he took advantage of the shelves of books around him and resorted to the local branch of the city's public library for the literary genres missing from his father's store, notably fiction and science books.

Of the four Trench children, Gregory was the only one to pursue a college education, and in 1979 he graduated top of his class from the Northwestern University School of Medicine. For the next fifteen years, life followed its predictable course for a brilliant young doctor in an affluent society: he completed his internship and specialization—in orthopedics—joined the staff of a private clinic, got married—to Terry Blum, an accountant and financial advisor—and bought a magnificent house in the upper-class Evanston suburb, just north of Chicago. As Dr. Trench's medical practice flourished, so did the couple's assets, with the help of Terry's shrewd investment decisions. It didn't take long for Gregory and Terry to become multimillionaire-rich.

Despite their wealth, during all those prosperous years the couple led a rather mundane existence. They had no children and didn't do much socializing; work was their favorite activity and pastime, punctuated with the occasional vacation or family gathering. As for religion, they no longer practiced the faith of their respective parents— Terry came from a Jewish family—or any other faith, for that matter. But that is not to say that Gregory, in particular, was totally indifferent to spiritual or religious questions. In fact, he had not lost his appetite for reading but had become more selective about the subject matter: he had developed a keen interest in the history of religion.

He was also a regular participant in an online forum where questions about the history, psychology, and philosophy of religion were discussed. "The Crucible" had been initiated by Peter Graham, a Canadian Anglican priest. His primary motivation for creating the forum was to fight religious intolerance and to foster a better under-standing among believers of different creeds, with some kind of rec-onciliation of all religions as the ultimate goal. He dreamed of a world free from sterile and divisive fights over which religion is the "true" one, a world in which worshippers of every faith would respect each other and equally honor the teachings and messages of Buddha, Moses, Jesus, Mohammed, or Guru Nanak.

People of all denominations together with atheists and skeptics and the inevitable crackpot logged onto the site, some disillusioned and pessimistic, others full of hope and compassion, all offering their opinion and advice on a wide variety of topics with the common thread of the notions of "God" and "religion."

There were those who would use the forum to mount provocative attacks against religious beliefs:

"Consider (a) 'My religion forbids me to kill (or to eat beef, or . . .)' and (b) 'My conscience forbids me to kill (or to eat beef, or . . .).' Is (a) more respectable than (b) just because 'religion' is mentioned? What's the difference between a 'religious' belief and a 'nonreligious' one if both are deeply and sincerely held? If a person believes that extraterrestrials visited the earth or that Americans never landed on the moon, why should those beliefs be taken less seriously than, say, believing in angels or that after death we go to heaven (or to hell)?

"Using the alleged 'religious' character of some beliefs to demand special treatment, privileges, or fiscal advantages is unfair and dis-criminatory. When will the artificial distinction between 'religious' and 'nonreligious' beliefs end and the rights of all believers be equally respected?"

Others would argue for tolerance toward different kinds of reli-gious practice, as advocated in "Song of the Hindu," a poem com-posed around 5000 BC by Karkarta Bharat, the supreme chief of the Hindu clan, and transferred by oral tradition into the written Vedas several thousand years later:

"Each man has his own stepping stones to reach the One-Supreme. . . .

"God's grace is withdrawn from no one; not even from those who have chosen to withdraw from God's grace.

"Why does it matter what idols they worship, or what images they bow to, so long as their conduct remains pure.

"There can be no compulsion; each man must be free to choose the path to his gods.

"A Hindu may worship Agni (fire) and ignore all other deities. Do we deny that he is a Hindu?

Another may worship God through an idol of his choosing. Do we deny that he is a Hindu?

Yet another will find God everywhere and not in any image or idol. Is he not a Hindu?

"How can a scheme of salvation be limited to a single view of God's nature and worship?

"Isn't God an all-loving universal God?"

There were echoes of Reverend Graham's own appeal against dogmatic barriers in this 7,000-year-old poignant plea for freedom of adoration. Sometimes, the convergence of views across the millennia can be truly astonishing.

A posting by a certain Tom Riley caught Trench's eye one evening, late in the spring of 1994. It vaunted the merits of Neo-Pythagoreanism, an ancient syncretistic religion—that is, one combining different religious and philosophical beliefs—seeking to interpret the world in terms of numbers and arithmetical relationships.

"*The origins of Neo-Pythagoreanism,*" Riley wrote, "*can be traced back to a school of philosophy based on the teachings of the famous Greek philosopher and mathematician Pythagoras that became prominent in Alexandria in the 1st century AD. Number One, or Monad, denoted to the Neo-Pythagoreans the principle of Unity, Identity, Equality, and conservation of the Universe, which results from persistence in Sameness. Number Two, or Dyad, signified the dual principle of Diversity and Inequality, of everything that is divisible or mutable, existing at one time in one way and at another time in another way. Similar reasons applied to their use of other numbers.*

Numbers could also be arranged in geometrical shapes, the most perfect of which is the Tetraktys (=Quaternary) consisting of the first four integers disposed in a triangle of ten points, where the number four is represented by all three sides of the equilateral triangle.

In this context, one represents the point; two represents the line; three, the surface, and four represents the tetrahedron, the first three-dimensional figure.

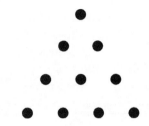

The Tetraktys

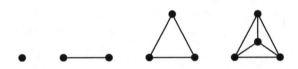

Hence, in the realm of space the Tetraktys represents the continuity linking the dimensionless point with the manifestation of the first body, while in the realm of music the Tetraktys contains the mathematical ratios that underlie the harmony of the musical scale: 1:2, the octave; 2:3, the perfect fifth; and 3:4, the perfect fourth. These ratios are represented by successive pairs of lines beginning from any vertex. For the Neo-Pythagoreans, then, the Tetraktys is the perfect symbol for the numerical-musical order of the cosmos."

The text explained that for the Neo-Pythagoreans there was a fundamental distinction between the soul and the body. Their religion was a purely contemplative one; they sought harmony, wisdom, and understanding for themselves and did not care about making converts or changing the world, "*and in this sense Neo-Pythagoreanism is an inherently 'pacific' and tolerant religion,*" Riley wrote, "*respectful of other faiths or beliefs.*" God must be worshipped spiritually by prayer and the will to be good, not in outward action. The soul must be freed from its material envelope by an ascetic way of life. Bodily pleasures and sensuous impulses must be forgone as detrimental to the spiritual purity of the soul.

"*The ancient Pythagoreans,*" Riley concluded, "*preached the virtues of a vegetarian diet and were against the killing of animals and eating*

*animal flesh. The current interest in vegetarianism and a humane at-
titude toward animals have their roots in Pythagoras' teachings. Vol-
taire described Pythagoreanism as 'the only religion in the world that
was able to make the horror of murder into a filial piety and a religious
feeling.' Actually, before the term 'vegetarian' was coined around 1842,
'Pythagorean' was the common name for those who abstained from eating meat."*

The mysterious Neo-Pythagoreans struck a chord in Trench, and
he wanted to learn more about the rites and tenets of their ancient
religion. A Google search turned up 412 results on "Neopythagorean-
ism," most of them short passages from academic articles or entries
from various encyclopedias repeating essentially the same thing:

> Very little is known of the members of this school. The Neo-
> Pythagoreans had no priests or leaders other than Pythagoras
> himself who, out of reverence, was never named. They referred to
> him as "the Man" or "the inventor of the Tetraktys." A sect of Neo-
> Pythagoreans was founded in the first century AD by Roman aris-
> tocrats who, true to their obsession with secrecy, literally went
> underground to practice their beliefs. In what is now the historic
> center of Rome, they built a subterranean basilica, the ruins of
> which were discovered in 1917. The Roman sect disappeared in the
> third century, but similar groups are known to have existed at dif-
> ferent times and places throughout history.

Not entirely happy with the results of his search, Trench e-mailed
Riley, asking him whether he knew of any group of Neo-Pythagoreans in the United States. He suspected the man—if he was indeed a
man—of being himself a member of one such circle. Riley's brief
reply, a few days later, was inconclusive: "It's not impossible. How
would I know? These people are very secretive. Why do you ask?"
Good question. Trench wasn't sure of the answer, or perhaps he
didn't want to admit to himself that his interest in the enigmatic Pythagoras and his followers was becoming obsessive. During the
following months, he read all he could lay his hands on about Pre-
Socratic philosophy, the early Pythagoreans and, foremost, the life of
Pythagoras.

In the fall of 1994, Trench's tranquil and mostly happy existence was abruptly shattered. One evening, as she was walking back home after her jogging session, Terry was hit by a car that had run a red light. Rushed to the hospital, she never regained consciousness and was declared dead two hours later with an inconsolable Gregory at her bedside.

After the funeral, Trench took a three-week leave from the clinic. Terry's sudden death would have a dramatic effect on his behavior, or perhaps her absence only precipitated a process that was already in progress. In the space of a few weeks, his personality underwent a radical transformation. He became withdrawn and solitary, and avoided the company of even his closest friends. His parents invited him to spend some time with them in San Diego, where they had moved after his father retired, but he preferred to stay home reading, surfing the Internet, and working out in his well-equipped basement gym.

Going through his e-mail one morning he found a short note from Riley: "Still interested in ancient truths? I might be able to help. Tom Riley." An exchange of messages followed and a meeting was arranged.

Two days later, at three o'clock in the afternoon, Trench was sitting at one of the tables in the spacious Starbucks store at the intersection of Lake and LaSalle, in downtown Chicago, with a huge, steaming mug of black coffee in front of him. He didn't have to wait long. "Dr. Trench?" The female voice came from behind him. He turned around to face her. The woman smiled, amused at the surprised look in his eyes as she introduced herself: "I'm Gloria Sweeny, alias Tom Riley."

They shook hands. She was a diminutive older lady, well past her sixties, wearing a dark suit and holding on to a disproportionately large purse. She sat down and began to talk before Trench could ask for an explanation.

"I'm here on behalf of a group of people having . . . how shall I put it? Some common interests. We call ourselves 'The Beacon.'"

Strange, Trench thought. That was the name of his father's bookstore until he sold it when he retired.

"We believe," Ms. Sweeny went on, "that you might be interested in participating in a project of ours."

"I very well might, if only you could be a bit more specific." Trench spoke for the first time; he had trouble adjusting to the new Mr. Riley.

"Of course, but let me first ask you a question: Do you care about the future of the world? I mean, the future of the human race."

"Of course I care. Who wouldn't? But I don't see the point."

"You will soon. Let me ask you another question: Take a look at the state of the world. What do you see?" Before he had time to reply, she answered her own question: "War, widespread poverty, population explosion, rampant epidemics, religious fanaticism, nuclear irresponsibility, air, soil, and water pollution, fish stocks and cultivable land depletion, deforestation, global warming. Not a very pretty picture. And who do you think will be able to fix the mess we're in? National governments? The UN? The multinational corporations? The scientists? The power of prayer? Certainly not the Church, whose business is to save souls, not the planet. And even less can we count on our typical politicians, short-sighted and controlled by the special interest groups who got them elected, when not downright corrupt."

She painted a rather gloomy picture. Trench still couldn't see what the woman was driving at. He had come expecting to learn more about the ancient Pythagoreans and their cult and was now beginning to wonder whether he had wasted his time. Too polite to just get up and leave, he resigned himself to hear the lady out. As if she had read his thoughts, her next question was more to the point.

"What do you know about Pythagoras, I mean, the kind of man he was?"

"Well, from what I've read, he was an extraordinary man, a man of many talents, a great thinker, and a brilliant mathematician. He was above all a spiritual leader—and also a political one, you may say—with a superior mind and a magnetic personality. The personage fascinates me, and I wish we had someone of his caliber around today."

"I couldn't agree more! But why settle for less, why settle for 'someone of Pythagoras' caliber?'" There was a pause while Trench tried to figure out what she meant.

"I'm not following you. . . ."

"Dr. Trench, what would you say if I told you that, at this very moment, Pythagoras may be among us, living somewhere in the world? And I don't mean someone *like* Pythagoras, but the Man himself."

"Now you've completely lost me. Is this some kind of joke?"

"I've never been more serious; but I can understand your bewilderment. I'm going to tell you a story. Please listen without interrupting me until I finish. I'll be glad to answer any questions later."

Trench took a sip of lukewarm coffee. The woman was searching for something in her purse. Her left hand came out holding a leather-covered flask, and he noticed that she was wearing a silver ring with a white stone. She unscrewed the top, took a long gulp, replaced the top, and quickly put the flask back in her purse. Trench had watched the entire operation in silence. When their eyes met, she smiled, said "My medicine," and began her story.

"As you know, Pythagoras lived in the sixth century BC. What we know about him and his school was transmitted to us mostly through the writings of Greek historians. By some accounts, Pythagoras was a demigod endowed with supernatural powers, son of the god Apollo and the human Phytais. According to other sources, his father was Mnesarchus, a tradesman from the Island of Samos. While Mnesarchus was at Delphi on a business trip, he was told by the oracle that his soon-to-be-born son would 'surpass all men who had ever lived in beauty and wisdom, and that he would be of the greatest benefit to the human race.'

"Unfortunately, that happened 2,500 years ago, and it is today that the human race could most benefit from Pythagoras' wisdom and supernatural powers."

Trench was about to ask something but remembered he had agreed not to interrupt her. The woman went on.

"Now, the works of Greek historians and philosophers that are our main source on Pythagoras were written several centuries after his death. Some of these works hint at the existence of an older and more reliable set of documents on the Man and his philosophy, a first-hand account written by some of his disciples shortly after Pythagoras' death and meant to preserve their Master's teachings, but only among the closed circle of the Pythagoreans and their descendants."

She paused and moistened her lips. Trench sensed that the woman was approaching the story's high point.

"One of the members of our group, I shall call him Mr. S, traveled extensively throughout Europe and the Near and Middle East, visiting

libraries, archives, and antiquarian book dealers, including underground traders and private collectors, in search of some trace of the
documents supposedly written by Pythagoras' disciples. Although he
didn't find any of these, he did discover compelling evidence, such as
verbatim quotations from very reliable sources, pointing to an extraordinary fact: around the years of the 680th Olympiad, Pythagoras
would be reincarnated to 'combat the evil.' Now, the years of the 680th
Olympiad in the ancient Greek calendar correspond to the middle of
the twentieth century. . . ."

She looked intently at Trench. If she had expected some reaction
from him, the doctor disappointed her.

"Mr. S then founded The Beacon," she went on, "a select society of
worshippers of Pythagoras whose main purpose is to find the Man reincarnate, the only person who can save the human race from extinction and prevent the disappearance of intelligent life from this planet.
He'll become our Master, Dr. Trench. He will lead and we shall follow."

Trench thought: Her "society" sounds to me very much like a sect
of Neo-Pythagoreans.

"But there's more," she continued. "Mr. S' research also revealed
that the clues that would allow us to recognize Pythagoras were contained in a parchment or papyrus written by one of his followers, or
maybe by the Master himself. The original document has almost certainly been lost or destroyed, but some sources point to the existence
of a well-preserved copy waiting to be found by someone who would
know where to look and what to look for.

"Would you be interested in joining our group and help us find
Pythagoras, Dr. Trench? We need all the help we can get—from people
we trust, of course."

It was a direct question. Trench was beginning to assess all its implications when Ms. Sweeny spoke again:

"Naturally, like any other potential member you would have to un-
dergo an acceptance process: answer some questions, disclose some
facts about yourself, that kind of thing. We need to find out how serious your intentions really are. And we would expect you to devote all
your time and energies to our common cause."

A strange thing then happened to Trench. A few moments earlier
he would have regarded the woman's offer to become a member of

her mysterious circle as preposterous: How could she dare imagine he would abandon a successful and lucrative medical practice to join a bunch of religious freaks believing in reincarnation?

But the next moment it struck him, with the force of a revelation, that accepting the woman's offer was the right thing to do, that for some obscure but compelling reason it was an opportunity not to be missed. As if in a dream, he heard himself saying, "Just tell me what I have to do to join The Beacon."

At the end of his three-week leave, Trench did not return to work. When his secretary called to find out what was going on, she listened in disbelief as he spoke in a calm but resolute voice: "I'm not coming back, Sarah. Please tell Dr. Thompson that my letter of resignation will be on his desk by tomorrow. My attorney will take care of the loose ends and the legal stuff. I'm going to take a trip and won't be around for a while. Don't worry, I'm alright. I know what I'm doing. Thanks for everything." And he hung up.

Chapter 15

THE TEAM

Three years later, Trench had become a Companion, the second-highest grade in the Order of The Beacon, a Neo-Pythagorean sect whose activities he had generously helped to fund.

The Order operated in the most secretive fashion, so that practically nothing was known about it outside the close-knit circle of its fifty or so members. These, who called themselves "fellows," were generally wealthy and influential professionals and businesspeople, some 20 percent of them women. In Pythagoras' doctrines, which they interpreted with an emphasis on their esoteric aspects, they had found a kind of spiritual fulfillment that mainstream religions had not provided. Such was their veneration for the famous philosopher that they worshipped him as a deity.

The Beacon had been founded in 1979 by a mysterious personage known to the other members only as Mr. S, who was convinced that Pythagoras would be reincarnated around the middle of the twentieth century. And so, not only did the members of the Order adore the spirit of Pythagoras, they also anxiously awaited his return among the living, the way followers of other religions wait for the coming of a messiah.

New members were admitted following a proposition from two fellows, and only after a thorough scrutiny of the candidate's background, personality, lifestyle, and motivation had established that certain strict criteria were met. On acceptance, the new fellow had to take an oath of loyalty and vow to financially support the Order and its search for Pythagoras reincarnate.

Fellows led their normal lives in society, but on the tenth day of each month they met at the Temple, the Order's headquarters located in the outskirts of Chicago, to perform a ceremonial ritual where they bonded together in harmony with the universe. By reaching a trance-like state, they sought to purify their souls and receive wisdom and understanding from the Higher Powers. It was also during these meetings that new members were received into the Order.

Trench still vividly remembered his induction ceremony. They were gathered in the Temple's Hall of Truth, a large, wood-paneled room with high ceiling and narrow windows and a podium at one end. A polished metal sphere, twice the size of a beach ball, hung suspended above the red-carpeted podium, reflecting the light from recessed lamps in the ceiling. The smoke rising in twin columns from two bronze burners at the front drifted through the room, filling the air with the sweet smell of incense. On the wall behind the podium was a screen with curtains drawn back on each side and showing the image of a five-pointed star. As the ceremony proceeded, a series of other images would appear on the screen at regular intervals.

The five members of the Council of High Companions, the Order's highest authority, were sitting in a semicircle on high-backed chairs at the center of the podium, facing the audience of fellows attending the ceremony. In front of the Council members was a round pedestal table with granite top and a red velvet runner on which sat a single object: a silver ring with a white stone.

All those present, including Trench, were wearing a hoodless white ceremonial robe with a silver belt tied around the waist. An elaborate geometric figure, embroidered in silver thread at about breast level, identified the bearer of the robe according to his or her grade, from the lowest, Fellow, to the highest, High Companion of the Order.

Trench stood before the Council with his back to the audience while the Elder—the Council's eldest member, a balding man in his sixties with a square jaw and tufty eyebrows—read to him the duties and obligations of a Fellow of the Order, among these the observance of absolute secrecy in all matters regarding the Order. He then took the oath of allegiance, which began "I, Gregory James Trench, solemnly swear by he who revealed to our soul the sacred Tetraktys. . . ."

While he spoke, the screen showed an image of ten points arranged in the shape of an equilateral triangle.

When he finished, he sat on a low stool near the pedestal table, facing the assembly of fellows. Someone brought a shallow silver bowl filled with water and a white linen towel. He washed his hands—the cleansing prior to his receiving the symbol of the Order.

Trench got up and was joined by the Elder, who picked up the ring from the table, took Trench's left hand, and with a deliberate gesture put the ring on his fourth finger. He then embraced Trench briefly as he said "Welcome among us, brother," and kissed him lightly on both cheeks as a token of friendship—friendly relationships were highly valued among the early Pythagoreans. One by one, the other members of the Council and the fellows in the assembly did likewise, after lining up in an orderly and slow procession that entered the podium from the left and exited it from the opposite side. The induction ceremony over, Trench sat in his assigned place among the fellows.

A reading of some of Pythagoras' precepts came next. After hearing each maxim, they all responded in a chorus "Yes, I will" or "So the Master said," depending on its meaning. At the end of the reading, all heads bowed in meditation and several minutes of total silence followed.

When the music began, they all stood up. The enveloping sound of a string instrument and of drums beating in the background filled the room; softly, barely audible at first but slowly growing louder. Trench looked around at the assembly. Everybody was standing, hands joined as if in prayer and eyes fixed on the screen at the front, which was now showing the image of a glowing and slowly rotating pyramid. As the minutes passed and the volume of the music increased, many began to speak in a strange language or perhaps only make sounds—Trench couldn't tell—as they kept staring at the hypnotic screen. Someone near him bellowed like a bull in dire pain—a terrifying sound. Instinctively, Trench closed his eyes and pressed his hands over his ears, but he could not shut out the frantic beat of drums, now resounding like a subterranean thunder threatening to burst through. Those assembled in the Hall of Truth were about to reach the climax: a trance state, a state of bliss that would put them in direct communion with the Master and other divinities. But only the purified, the initiated in the secrets of Number, could expect to communicate with the gods. Trench was not one of them, not this time, not yet.

In the three years that had passed since Trench joined the Order, the search for the manuscript that would reveal the clues to finding Pythagoras reincarnate had gone on without success. The possibility that the only extant copy of the precious document could be gathering dust on the shelf of some private collector's library, ignored and inaccessible, did not dampen the sect's determination or undermine their faith in the final outcome. They were driven by the profound conviction that, not only was Pythagoras' reincarnation alive somewhere in the world, but he would eventually be found and become their spiritual leader, and accomplish many marvelous things that would benefit all of humankind.

Several promising leads turned out to be dead ends, including a forged parchment filled with mathematical formulas—its owner pretended it was an autograph in Pythagoras' own hand, ignoring the fact that parchment was only invented in the second century BC, more than three hundred years after the philosopher's death. Despite all the uncertainties and disappointments, the search went on.

In early December 1997, a fax from London set in motion a series of events that would culminate in a breakthrough. Trench, now in charge of coordinating the search operation, had built a network of "agents" who were on the lookout for anything resembling the kind of document he was seeking. His network was made up mostly of collectors and antique book dealers in various countries, including some unscrupulous individuals who operated in the black market of smuggled or stolen items. One of these shady traders, code-named "Emerald," had just bought a fragment of a medieval book on parchment containing what appeared to be artistic decorations of a Greek text, but which on closer inspection also revealed some intriguing mathematical and geometrical symbols. Just in case, Emerald decided to fax pictures of the pages to Trench. "Thought you'd like to have a look at this," read Emerald's laconic message on the cover page. Eight other pages followed. The images were of poor quality but good enough for Trench to see something that aroused his interest. He immediately e-mailed back, instructing Emerald to hold on to the book until further notice.

The pentagram

What caught Trench's eye was a small geometrical symbol that appeared on two of the pages: a five-pointed star drawn inside two concentric circles.

The pentagram, also known as the triple interwoven triangle or the star-pentagon, was long believed to possess magical powers. In Babylonia and ancient Greece it was worn by people practicing pagan faiths. Pythagoras' disciples saw in the five-pointed star a portrayal of mathematical perfection related to the so-called divine proportion or golden mean. They wore it as a sign of recognition among the members of the school and as a symbol of inner health.

Trench was particularly excited by the fact that the pentagrams were drawn with two points up and the letters in its vertices formed the Greek word for "health," precisely the way the early Pythagoreans drew the symbol. Could this be what he was looking for?

The eight pages were clearly part of a larger set, probably a book. Was the rest of the book, if it still existed, also up for sale? Emerald had not been very forthcoming about the circumstances surrounding the purchase of the parchment leaves, but with some reading between the lines Trench was able to get the picture: Emerald's "client" was a small-time thief "doing the rooms" of some London hotel who had come across the book fragment by chance during one of his rounds. Nothing was known about its owner except that he or she was probably a foreigner. Whatever the case, the ancient parchment leaves certainly deserved a closer look.

One week later, the order's Council of High Companions held a meeting to discuss the course of action following the acquisition, on Trench's recommendation, of the medieval book for the sum of 20,000 pounds.

A thorough examination of the pages had convinced them that their elaborate mixture of artistic work, geometric diagrams, and mathematical symbols—not to mention an enigmatic short poem at the end—concealed a message, possibly a piece of the puzzle leading to the reincarnated Pythagoras. As to the nature of the message, they could only speculate: it could be a reference to a place or to another text, perhaps even some mathematical discovery to be revealed only to the initiated but irrelevant for the purpose of identifying the living Pythagoras.

The Council then decided that they needed help to decode the hidden message and charged Trench with the task of hiring someone with the necessary skills. The chosen person would be the fourth member of a team whose mission would be finding Pythagoras' reincarnation and bringing him before the Council.

One afternoon at the beginning of January 1998, a car arrived at the Order of the Beacon's headquarters, an imposing nineteenth-century mansion situated in a densely wooded area north of the city. The property was surrounded by a high wall, with access through a security gate. The driver of the car was Leonard Richter, and his only passenger was Jule Davidson, the latest—and final—member of the search team.

After being cleared by the security agent at the gate, Richter parked the car at the back of the house and they entered the building through a side door. Inside, the air was warm and dry, a welcome contrast to the cold wind that had blown over them on their short walk from the parking lot. They crossed a large vestibule with tiled floor and left their coats in the adjacent cloak room. "You may leave your bag here," Richter said to Jule, who was to stay overnight at the Temple, "someone will take it to your room."

They went up a wide, carpeted staircase to the first floor and walked along a dimly lit hallway toward a glass-paneled door. Richter knocked before opening the door and leading Jule into what appeared to be a library. Bookshelves covered every wall except the one in front of them, where two large valence windows with heavy draperies drawn back let in the fading afternoon light. The air in the room was warm and had the pleasant scent of wood smoke. Gregory Trench was sitting at a big cherry wood desk by the fireplace. He looked at them over the rim of his reading glasses. "Good afternoon, gentlemen. Please come in," he said, getting up and coming forward to meet them.

Richter introduced Jule to Trench: "This is Doctor Jule Davidson, the new member of our team." Trench fixed Jule straight in the eyes—the mathematician was shorter than him by about a foot—before extending his hand and saying with an even voice, "I'm Gregory Trench. Welcome on board, Doctor Davidson." Jule smiled politely and extended his hand in return. "Glad to be part of the team," he said while they shook hands, and he added: "I look forward to learning a bit more about the nature of my work."

"Of course," said Trench without emphasis, still looking Jule deep in the eye, studying him, as if he were searching for something confirming that the man was up to the task. He finally invited his guest to sit down. "Please make yourself comfortable, Dr. Davidson." Jule sat down on a leather armchair facing Trench, who settled on a red and gold damask sofa, its colors matching those of the draperies behind it. Richter excused himself and left the room as Trench began briefing Jule about the operation and his role in it.

"As you know from Mr. Richter, we're a group of believers in the principles of Neo-Pythagoreanism, an ancient syncretistic religion based on the teachings of the sixth-century BC philosopher and mathematician Pythagoras of Samos, the wisest man who ever lived. He traveled extensively in Greece, Egypt, and the East, and the knowledge he acquired drew from the wisdom of all those peoples, which was considerable. To prevent this knowledge and the science and philosophy he had taken such great pains to elaborate from being lost, he created a brotherhood of disciples for their preservation and transmission, and thus became one of the chief benefactors of mankind. As heirs of that millenary tradition, our duty is to carry on this noble and sacred mission."

He paused and moved forward in his seat a little. When he spoke again, there was a dramatic, almost threatening intensity in his words.

"We have reason to believe, and you're not to question our conviction or demand any proof or justification whatsoever, that in this very moment the reincarnation of our Master is living somewhere in the world. Your job is to help us find him."

Jule would be working with Professor Hirsch, an expert on ancient Greece, whom he would meet the following day. Their task would be to identify Pythagoras reincarnate. They would report to Richter and, in his absence, to Trench. Everything related to the operation, the Temple, and its occupants should remain strictly confidential, he warned him.

Trench also mentioned there were two other members on the search team, without further explanations. In the weeks ahead, Jule would learn more about them and their role in the operation. Rocky was a mountain of a man, strong as a bull; a one-time professional wrestler who for a while lived on the wrong side of the law. He had been "adopted" by Trench, after saving Trench's life one night during a mugging attempt that was about to turn ugly. Thanks to Trench he had "seen the light" and his life had changed; he would do anything for him, even kill. As for Houdini, a tall, wiry man in his thirties who spoke little, he was a mechanical and computer wizard. He could pick a lock, open a safe, or break into a computer as easily as he breathed. Of a solitary and introspective nature, he lived alone and designed video games for a living. Once Jule and the Professor had discovered the identity of the living Pythagoras, Rocky and Houdini would set out to find him and convince him—or, if necessary, compel him—to come to the Temple.

Later that evening, when he was alone in his sleeping quarters, a sparsely furnished room with no windows, Jule began having second thoughts about his hasty acceptance of Richter's offer. He had a problem with reincarnation—he simply did not believe in it—and with the thinly veiled fanaticism he sensed in the members of the sect. It's too late to turn back, he told himself. He was no quitter; he decided he would do his best and play the game to the end.

Chapter 16

THE HUNT

She had been introduced to Jule by Trench as "Professor Hirsch, an authority on ancient Greece." And, after the customary exchange of polite greetings, Trench had added: "Professor Hirsch is presently writing a book about Pythagoras. She will be your resource person on him and his school."

Her full name was Laura Eva Hirsch. She was born in the former German Democratic Republic, where she studied Greek history and mythology before escaping to the United States as a political refugee with the help of a diplomat friend. In 1983, she received her PhD in classical studies from the University of Illinois. She stayed on at her alma mater as a teacher, moving quickly through the ranks until she became a full professor in 1989. Now in her late forties, she was a tall, slender woman with an intelligent if not attractive face, her dark brown hair and black eyes accentuating the whiteness of her skin.

It was Saturday morning. Jule and the professor were alone in a large room with tall windows overlooking the park at the back of the house. A blanket of fresh snow extended into the distance, interrupted only at the far end by a dark wall of pine trees. High above the trees, a few small clouds floated against the pale blue sky like cotton islands in a vast, calm sea. Jule had had dinner and breakfast served in his room by a silent and unsmiling old lady, so this was his first contact with someone other than Trench or Leonard Richter.

The professor spoke first. "I'll be assisting you with anything you wish to know about Pythagoras and his disciples," she said to Jule, who noticed that she spoke with a slight German accent. "Think of me as a kind of human Google," she added with a smile.

"I'll need all the help I can get," replied Jule, smiling back, happy to learn that the professor didn't seem to take herself too seriously. "Why don't you begin with an introduction to the man and his doctrines, anything you would care to tell me for starters?" he suggested, flipping open his notebook.

"All right," she said, with an approving nod. She sat back in her leather chair and, after a few moments' reflection, began her lecture.

"In his time, Pythagoras was revered as a man of exceptional wisdom, almost as a demigod. But today he is remembered primarily for the famous theorem named after him, although the result was known to the Babylonian and Indian civilizations centuries before his birth. I remember learning it by rote as a youngster—in German, of course, my mother tongue: 'In a right-angled triangle, the square of the hypotenuse equals the sum of the squares of the two other sides.'"

Jule half-raised his hand, like a student trying to get the teacher's attention. Laura Hirsch smiled and said, "Yes?"

"You just said that Pythagoras' famous result was known to more ancient civilizations. Do you mean then that Pythagoras' theorem is not really Pythagoras'?"

"No, I didn't mean that, but the question of 'ownership' is a difficult one, especially when the claimants have been dead for thousands of years and what few records remain are open to different interpretations. Take, for example, a fragment of a Babylonian clay tablet dating from 1700 BC and kept at some museum—in New York, I believe. It consists of four columns of numbers, written in cuneiform script and in the system of numeration the Babylonians used. According to modern scholars, it is essentially a table of so-called Pythagorean triples, that is, numbers a, b, and c such that $a^2 = b^2 + c^2$—actually, only the columns corresponding to the values of a and b appear on the extant piece of the tablet. One may then assume that the Babylonians knew the famous theorem we attribute to Pythagoras. But did they really?"

"I see your point," said Jule. "It's not absolutely clear that those numerical equations arose from their knowledge of the geometry of right-angled triangles; I mean, that they were actually listing the relationship between the hypotenuse, a, and the sides, b, c of right-angled triangles. They could have been merely showing examples of an arithmetical curiosity: square numbers that are sums of two square numbers."

"It's possible. But although the table doesn't explicitly mention any geometrical relationship, the headings of the columns for a and b read 'square-side of the diagonal' and 'square-side of the width,' respectively, so they could well have known the theorem after all.

"Besides, there is another Babylonian tablet, with a diagram and three numbers written on it, which might suggest the Babylonians were aware of a visual proof of the theorem for isosceles triangles.

"As for Indian mathematics, the evidence is perhaps more compelling. Baudhâyana was an Indian priest and mathematician who lived around 800–600 BC. He probably was not interested in mathematics for its own sake but for its use in the construction of altars needed for sacrifices to the gods and other religious rites. For these sacrifices to be successful and the gods to grant the people's wishes of good health, abundance of food, and so forth, the altar had to be built according to very precise measurements. Baudhâyana's *Sulba Sutra* or 'Rule of Chords,' written in Sanskrit and without using any mathematical symbols, is a collection of mathematical results and geometric constructions stated without proof. Among these, we find the following statement." She searched through her papers. "Here we are: *A rope stretched along the length of the diagonal produces an area which the vertical and horizontal sides make together.*"

"That's amazing!" Jule cried out in admiration. "It's essentially the general form of what we call the Pythagorean theorem. I read that as a young man Pythagoras traveled to India. Could he possibly have learned about the result over there?"

The professor answered in an indirect way: "It's not certain that Pythagoras traveled as far as India, although it is well established that he visited Egypt. In my opinion, such a fundamental mathematical result, especially one with obvious practical applications—carpenters still use it today for making square corners—would have been discovered sooner or later in any advanced civilization, so it may well have more than one discoverer.

"Actually," interrupted Jule, "what carpenters really use is not the theorem itself but its converse: *if* $a^2 + b^2 = c^2$, *then* the triangle is right-angled—a distinction that is lost on most people."

"I know, I was one of those people," Ms. Hirsch confessed with a nod. "The Chinese too have claimed the paternity of the theorem," she went on. "There exists a very ancient Chinese mathematical text* dating at least from the time of Confucius in the sixth century BC, if not earlier, with a drawing of a square askew within a larger gridded one, and a commentary accompanying it. Together, they provide a visual proof of the theorem for the 3, 4, 5 triangle."

"Why then did the theorem become associated with the name of Pythagoras?"

"The truth is we really don't know. The earliest reliable source is Plutarch, a Greek writer from the first century AD, who reported that according to a certain Apollodorus the logician, Pythagoras offered a splendid sacrifice of oxen to celebrate his discovery that 'the square of the hypotenuse of a right-angled triangle was equal to the squares of the sides containing the right angle,' and that the event inspired an epigram:

When the great Samian sage his noble problem found
A hundred oxen with their life-blood dyed the ground.

However, such a 'celebration' appears most unlikely, given the Pythagoreans' convictions against killing animals and eating animal flesh."

"How about *the proof* of the theorem, then?" asked Jule, attacking the question from a different angle. "There exist hundreds of different arguments establishing the validity of the proposition, but how did Pythagoras prove it—if indeed he did?"

"I know there are hundreds of proofs,"† said Ms. Hirsch before answering Jule's question, "including one in the nineteenth century from a certain James A. Garfield, who would later become president of the United States. But how Pythagoras might have convinced himself of the truth of his proposition is something we don't know. In fact, we must first elucidate the notion of what constituted in his time a mathematical proof."

"The famous British mathematician Michael Atiyah once said that proof is the glue that holds mathematics together," observed Jule.

*The *Zhou bi suan jing.*
†See appendix 4 for a visual proof of the theorem.

"But the concept of proof evolved over time. In 1988, the nonexistence of a projective plane of order 10, a two-century-old question in geometry, was finally settled—in the negative—using a supercomputer to carry out certain parts of the proof. Problem was, only a computer could check those parts of the proof. As the *New York Times* put it: 'Is a maths proof a proof if no one can check it?'"

"Let's get back to the Greeks, if you will." Eva Hirsch hadn't appreciated the digression into twentieth-century mathematics. She searched on her laptop for a few moments. "Here it is. These are excerpts from a book on the beginnings of Greek mathematics. Read for yourself." She handed him the laptop.

> The most substantial difference between Greek and Oriental sciences is that the former is an ingenious system of knowledge built up according to the method of logical deduction, whereas the latter is nothing more than a collection of instructions and rules of thumb, often accompanied by *examples*, having to do with how some particular mathematical tasks are to be carried out. [...]
>
> The mathematical term for *proof* and *to prove* is the Greek verb δεικνυμι, and it seems most likely that right from the start this verb meant "to point out" in both a figurative and a literal sense, and hence also "to explain". [...] Some scholars, without making any reference to this verb, have stressed the importance of visual evidence in early Greek mathematics. This suggests the idea that the earliest "proofs" may have involved some kind of "pointing out" or "*making visible*" of the facts; in other words, δεικνυμι may have become a technical term for "proof" in mathematics because "to prove" meant originally "to make the truth (or falsity) of a mathematical statement visible in some way" [...].
>
> Here is an example of mathematics teaching in antiquity that seems to support our conjecture. It is the passage from Plato's *Meno* where Socrates asks a slave how the area of a square with sides two units long can be doubled without altering its shape. He then draws a diagram [see (1) in the figure] *showing* the square that is to be doubled. After the slave answers that the square will perhaps be doubled by doubling the length of its sides, Socrates draws a second diagram (2) to *show* that a square whose sides are twice as long

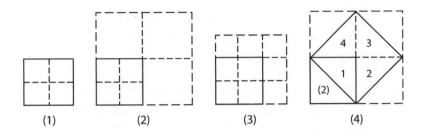

(1) (2) (3) (4)

as those of the original one, has an area four times its size. He then goes on to draw a third diagram (3) *showing* that a square whose sides are three units long, has an area of nine square units, and hence cannot be the double of the original one. Finally in his fourth diagram (4) he *shows* that the square on the diagonal has exactly twice the original area.

I think this passage from Plato provides an excellent illustration of the way in which statements were verified at an early stage in the development of Greek mathematics, i.e., it tells us how propositions were "proved" in the archaic sense of this word. The proposition that has been illustrated and "proved" above, can be formulated as follows: "A square constructed on the diagonal of another has twice the other's area." [...]

On the other hand, the proofs in Euclid's *Elements* [some two centuries later] are already in a polished form. [...] He is less interested in making his propositions visually evident than in convincing the reader of their truth by a train of abstract arguments.

Jule finished reading and remained silent, staring at the computer screen. A minute went by before Eva Hirsch finally broke the silence: "So? What do you think?"

Jule answered with a question: "Was Socrates a contemporary of Pythagoras?"

"No, he was born some seventy years after Pythagoras' death. Why do you ask?"

"Because his solution of the problem is also a very simple proof of Pythagoras' theorem, albeit in the particular case of an isosceles triangle."

"How's that?" She seemed to have been caught by surprise.

"Come and take a look at Socrates' fourth diagram." She moved over beside him.

"The diagonal of the square is also the hypotenuse of a right-angled triangle whose two other sides are the sides of the square," Jule explained, as his finger traced out the triangle on the screen. "So the square on the diagonal is the square on the hypotenuse, and Socrates showed that its area is twice that of the original square, in other words, equal to the sum of the squares on the two other sides."

It only took the classics professor a moment to grasp Jule's argument. "I'm impressed," she said, turning her head to look at him.

Jule ignored the compliment. "I bet Pythagoras used a similar geometric construction to convince himself of the truth of the proposition in the general case. By the way, your Indian mathematician who stated the theorem eons ago—'a rope stretched along the diagonal . . .'— never mentioned a rectangle, so maybe his statement was only about the particular case of the square too."

"You're right—and not the first one to have raised the question. But let's move on. I'm supposed to be telling you all about Pythagoras and his school and I've hardly begun."

She went back to her chair and resumed her lecture.

"For Pythagoras, the intellect was the most important human faculty, for it can lead to a form of knowledge that surpasses any other in depth and certainty. 'Let reason, the gift divine, be thy highest guide,' he taught. His vision of nature was based on the principle that numbers rule the world, and is often condensed in the dictum 'All is number.' He seemed to have reached this sweeping conclusion after observing that musical intervals pleasing to the ear, such as octaves, fifths, and fourths, as they are now called, come from plucking two taut strings whose lengths are in the same ratio as two whole numbers—2 to 1, 3 to 2, and 4 to 3, respectively. Such a bold generalization may appear naïve to us, but what if the Pythagoreans had found evidence in support of their belief? What if they had discovered that the cosmos is a kind of infinite hologram? Then, just as a small fragment of a hologram contains all the features of the whole, so all the secrets of the universe may be hidden in any tiny speck of it—a crawling insect, a child's smile, or a vibrating string.

"Echoes of the Pythagorean tenet may still be found 2,000 years later in Galileo's belief that the Book of Nature is written in the language of mathematics. In fact, many famous scientists, from Kepler to Einstein, to name only two, shared Pythagoras' vision of a universe ruled by numbers and harmony, one whose inner workings could be grasped by uncovering the underlying mathematical relationships.

"Kepler's discovery at the beginning of the seventeenth century of the laws of planetary motion was guided by his firm belief in Pythagorean principles. For him, geometry had provided God with a model for the Creation; and the orbital speeds of planets were a reflection of consonant musical intervals, so that, in their periodic journey around the Sun, the heavenly motions produced the 'music of the spheres,' perceived not by the ear but by the intellect.

"As for Einstein, not only did he reject the probabilistic foundations of quantum mechanics on the basis that 'God does not play dice with the world,' but he also wrote that a scientist may count as a Platonist or Pythagorean if he considers the viewpoint of logical simplicity as an indispensable and effective research tool, a viewpoint that he had himself adopted.

"But after having discovered galaxies and genes, quantum phenomena and chaotic systems, and caught a glimpse at the complexity of biological processes, we now know that the universe is not as simple as 'all is number.' And yet . . .

"It is thanks to numbers that we can swiftly store, transmit, and reproduce not just information in various forms but also sounds and images. In a very real sense, a CD, DVD, digital picture, or movie *is* a sequence of numbers—the 0s and 1s of its binary code. Decisions affecting the fate of most individuals are based on statistics, a sophisticated form of numerical manipulation. Numbers are also used in subtle ways to control, mislead, or deceive, as when complex concepts are reduced to a single number, when 'intelligence' is identified with IQ, or the health of that intricate reality we call 'the economy' is expressed numerically as gross national product.

"Numbers alone might not rule the world of nature created by God, as Pythagoras believed—and in that sense he has been proven wrong—but they increasingly reign over the human-made world that is life in a modern society. That's Pythagoras' revenge."

"Is that also the title of your book?" asked Jule.

"Tentatively, yes," she replied, "but editors always have the last word on such matters."

Professor Hirsch went on with her lecture for a good three hours, occasionally reading from a thick file of notes or her computer screen and interrupted every once in a while by a question from Jule. They hardly noticed the old woman when she came in to leave a tray with coffee, sandwiches, and some fruit.

Trench dropped in at around four o'clock. He filled them in on the mechanics and other details of the operation known as "The Hunt": finding Pythagoras reincarnate and bringing him to the Temple. Jule and Laura would be in charge of the "finding" part of the undertaking (who and where is he), while Rocky and Houdini would be called in for the "bringing" phase, when—and if—Jule and Laura succeeded.

For the coming weeks—or months—they would stay at a hotel in the city and work at Richter's place in Highland Park, where an office had been equipped with the latest technology.

Their task was not an easy one, to say the least, not to mention the fact that they could very well be chasing a ghost if the story of Pythagoras' reincarnation turned out to be only a myth, a legend like so many other ancient tales. Blinded by their faith, the members of the Neo-Pythagorean sect hadn't even considered such a possibility—and Laura and Jule were not to contemplate it either, as Trench had made clear to them—so the next question was where to start.

Chapter 17

THE SYMBOL OF THE SERPENT

A week after Jule and Laura first met at the Temple, the search for Pythagoras reincarnate was in full swing. In the prevailing atmosphere of unwavering belief, Jule's initial skepticism regarding reincarnation had somewhat abated. He had proposed setting up some kind of bait message on the Internet and in newspapers around the world which might get a response from a potential candidate but would mean nothing to others. Trench, however, would have none of that, for fear of being swamped with calls from all kinds of impostors and freaks in search of attention or worse.

They began by looking at how other religions faced with a similar problem went about it and studied the case of the Tibetan Buddhists. After the death of a Dalai Lama, their supreme spiritual guide, a search team made up of monks and other Lamas is sent to look for his reincarnation. Their search may go on for several years. The thirteenth Dalai Lama, who died in 1933, had foreseen his reincarnation in a dream. He described a small house with peculiar gutters and an old poplar tree behind it, and the golden roof of a monastery in the distance. A place bearing a striking similarity to this description was found by the search party some four years later. A two-year-old boy called Tenzin Gyatso lived in the house with his mother. After the Lamas obtained her permission to conduct certain tests, they placed a number of objects in front of the boy; among these, there were some that had belonged to the previous Dalai Lama. They then asked Tenzin to pick out the items he liked. Familiarity with the possessions of the previous Dalai Lama is considered the main sign of reincarnation. When he finished, the boy had chosen all the objects that had belonged to the thirteenth Dalai Lama and none of the others. Tenzin also passed other tests and was declared the Dalai Lama reincarnate.

But the hunt for Pythagoras' reincarnation could not proceed along quite the same lines. To begin with, they had neither a description of a place nor tests to evaluate potential candidates; moreover, unlike the reincarnations of the Dalai Lama, which all occurred within a small geographical area and only a few years apart, there was no particular reason to expect Pythagoras to reappear, two-and-a-half millennia later, exactly where he had last lived—in what is now southern Italy. All they knew was that if Pythagoras was reincarnated in the middle of the twentieth century as predicted, they were looking for a man between thirty-five and forty-five years old—not really the kind of information that could significantly narrow down the search. Getting their hands on the manuscript with the clues for recognizing Pythagoras was therefore crucial.

The most promising lead they had was the medieval book, or rather fragment of a book, that Trench had bought from his London agent: four parchment sheets, eight pages in all, featuring a mixture of elaborate artistic patterns, figures, mathematical symbols, and what appeared to be a short poem, in Greek. Two of the sheets had been partially eaten by rodents, resulting in damage at the margins but without significant loss of content. The last page, however, infected with a purple mold, was practically illegible, even after having been carefully washed in an alkaline solution.

Just to make sure the sheets did not contain some hidden text or drawings, Jule had a spectroscopic analysis performed on them. The results came back negative.

Since the book was most likely a copy, it was possible that the copyist had added his own artistic contribution. In that case, some of the pages would be irrelevant as far as the cryptic message they were looking for was concerned. Whatever the case, Laura and Jule had the strong feeling, if not the conviction, that only two of the pages deserved their attention: those, the fourth and the sixth, on which the five-pointed star or pentagram drawn inside two concentric circles appeared—the symbol of recognition among the Pythagoreans.

They began by studying the sixth page, at the center of which was a short poem in Greek. The rest of the page was covered with arabesques, which were clearly there as mere embellishments. Laura had found the four verses vaguely familiar:

> He is a man of exceptional wisdom
> Versed in the secrets of Number;
> One who of all men
> Has the profoundest wealth of intellect.

Was this a reference to Pythagoras? He certainly fit that description, but so did several other ancient Greek philosophers and mathematicians.

Two days later, in the middle of the afternoon, Laura startled Jule with a loud "Eureka!"

"What is it?" he asked, turning around in his chair.

"Empedocles."

"Who?"

"The poem. It's a shorter version of some verses by Empedocles."

"Forgive my ignorance, but what verses, and who's Empedocles?"

"A Greek pre-Socratic poet and the father of rhetoric, according to Aristotle. As for the verses, I just found the reference in the ninth book of Timaeus' *Histories*:

> Among these was one in things sublimest skilled;
> A man of wondrous wisdom,
> His mind with all the wealth of learning filled.
> And when he extended all the powers of his intellect,
> All things existent, easily he viewed,
> As far as ten or twenty ages of the human race!

"Timaeus, a Greek historian, believes that Empedocles had Pythagoras in mind when he wrote his enigmatic verses, and he adds:

'The words 'sublimest things,' 'he surveyed all existent things,' 'the wealth of the mind,' and the like are indicative of Pythagoras' constitution of body, mind, seeing, hearing, and understanding, which was exquisite, and surpassingly accurate.'"

"This simply confirms what we already suspected," said Jule without much enthusiasm, "that these parchment pages have to do with

Pythagoras. But the question remains: Do they conceal a message? I can see none in the poem; it's just a reference to Pythagoras."

Laura, happy at having found the source of the poem, did not share Jule's thinly veiled pessimism: "Let's focus on the other enigmatic page, then. I'm sure it's the key to the mystery," she said, with more hope than conviction.

The fourth page contained a finely executed drawing. To properly view it, one had to rotate the book clockwise ninety degrees. The center of the drawing showed a serpentlike monster or dragon inside an almond-shaped lozenge. Symmetrically placed on each side of the lozenge was a pair of matching feminine figures dressed in the tunic and *pallium*, a Roman cloak of Greek origin, worn by sacred persons; one of them was holding a stick and the other a scroll. The figures seemed to grow out of a stylized lily flower at their feet and were flanked by two cornucopias.

Was this a reference to a scroll and to some high-ranking personage, possibly a priestess, guarding it? They didn't think so. The scroll was not sufficiently prominent for that; it was practically lost among many other elements in the image. Their attention turned rather to the drawing's focal point: the serpent at the center of the page.

They had no idea when the serpent could have been drawn. The best estimate they had obtained for the parchment's age was some eight hundred years—that is, dating it back to the thirteenth century—but the original drawing could have been as old as the first century AD, or even older. Jule counted on Laura's almost encyclopedic knowledge of antiquity for guidance.

"The serpent is an extremely ancient and universal symbol of knowledge and wisdom," she informed him. "Only in Christian times has it been endowed with demoniac attributes and used as a representation of evil."

Jule observed that the two feminine figures could be priestesses standing in reverence before the serpent and presenting the cornucopia as offerings to the sacred animal. "Was there any particular school or sect worshipping the serpent or using it as a central ritual element?" he asked.

"As a matter of fact, there were a number of them. Their members were known as Ophites, from the Greek *ophis*, 'serpent.' These various sects and groups flourished in the Roman Empire during the second century AD. They were convinced that they possessed a secret and mysterious knowledge, not accessible to those outside, and based not on reflection or scientific inquiry but on revelation.

"In particular, the serpent occupied a central place in the worship and the symbolism of the Naassenes, from the Hebrew 'nahash,' serpent. The members of this Gnostic sect regarded the spiritual dragon or serpent as a symbol of intelligence, the redeeming power through which Adam and Eve gained the all-important knowledge of good and evil and learned of the existence of a Supreme Being higher than Jehovah, their creator, who had withheld this knowledge from them."

"I see," said Jule in a reflective tone. He looked disappointed. After a few moments he spoke again, with a touch of frustration in his voice:

"I can't see any connection with the Pythagoreans. Their knowledge had not been revealed to them by any Supreme Being but came from the teachings of Pythagoras; and if they held something to be sacred it was certainly not a serpent but the Tetraktys, a triangular arrangement of ten points. I'm afraid we're on the wrong track."

Laura was not so pessimistic. "Not necessarily," she said. "Neo-Pythagorean sects existed as late as the second century AD, an age in which abstract philosophy and arid formalism had begun to pall. It is quite possible for a splinter group of Neo-Pythagoreans to have been lured by ophite rites while at the same time wanting to preserve the secrets of the past, including the one announcing Pythagoras' reincarnation. In that case . . . "

"Wait a second!" he interrupted her. There was a spark of excitement in his green eyes. "Maybe we're trying to read too much into the drawing," he said. "I mean, we're assuming that it carries some kind of coded message to be deciphered."

"Of course we are," Laura cut in. "Didn't we conclude that the presence of the pentagram was not fortuitous but an indication that the drawing had a purpose and a meaning related to Pythagoras?"

"I'm not questioning the fact that the drawing has a purpose related to Pythagoras," Jule calmly replied, "except that it doesn't have to have a meaning—at least not for us."

"But then . . ."

"What I'm trying to say is that perhaps the drawing is merely the reproduction of a similar image that exists somewhere, in a place where there *is* something related to Pythagoras. Never mind its meaning, its purpose is to call attention to that place."

Like sunlight bursting through the clouds, the plausibility of Jule's interpretation of the drawing suddenly hit Laura: "Of course!" she exclaimed with excitement, "I'm pretty sure you're right."

It was only a small step forward, but enough for them to feel they were making some progress—which until then had not been the case. Each day, they spent a good deal of time on the Internet, scanning the online international press for some story that would be relevant, even remotely, to the search. Not that they expected to find the heading: "Man Claims to Be the Ancient Greek Philosopher Pythagoras." There were some language limitations: Jule was fluent only in English and could read some French, while Laura also spoke German and Greek. However, they figured that if the news item was unusual enough it would also be carried in some English-language newspaper.

After their initial enthusiasm at the discovery that the drawing pointed to some particular place—they had no doubts about it—they realized that the road ahead could be long and difficult. The place they were looking for was in all probability some kind of ancient temple, and unless an archaeological expedition had dug it up at some point, they would have no reasonable hope of ever finding it. And even if the temple ruins had already been discovered, the painting or drawing on the wall or ceiling had to be in fairly good condition, good enough to be recognizable. Laura tried to find reasons to be optimistic: "When Evans dug up the ruins of the Palace of Minos, the legendary home of the Minotaur in Crete, he found many frescoes dating back to 1500 BC and most of them were relatively well-preserved. Who says we couldn't be just as lucky?"

They first had to identify the place. "Wouldn't it be nice," Laura thought, "if one could search the Internet 'by image' the way one does by word or phrase?" But search engines were not yet sophisticated enough for that, so only traditional methods could be employed. Laura would comb through the reports and archives of archaeological expeditions with the help of her numerous contacts in archaeological and historical circles. Since the sect of the Naassenes, which worshipped the serpent with particular fervor, was believed to have originated in Phrygia—a large region of Asia Minor corresponding approximately to modern Turkey—she decided to start by examining records of excavations in that area.

Chapter 18

A PROFESSIONAL JOB

Toward the end of January, Trench had news from one of his London agents: an ancient book believed to refer to Pythagoras was being offered at auction. Trench then decided to send Jule to London to look into the matter.

"I need you to travel to London and take a look at a book that's up for auction," he told Jule over the phone. "You'll be leaving tomorrow evening. I've already made plane and hotel reservations for you. Richter will give you an envelope with addresses and other details. Call me as soon as you know something."

"Sure"—that was all Jule had a chance to say before Trench ended the call with a polite "Have a good trip."

Two days later, after having examined the book, Jule called Trench, hardly containing his excitement: "I'm pretty sure our parchment sheets with the mysterious drawings were cut from this book, the owner probably expecting to fetch a higher price by selling them separately. The page size, appearance, and age of the parchment are very similar—not to mention the fact that both manuscripts have to do with the Pythagoreans and have surfaced at approximately the same time.

"The book is written in Arabic. According to the auction notice, it's most likely a twelfth- or thirteenth-century translation of a much older Greek manuscript by one of Pythagoras' disciples reporting the tragic circumstances of his master's death. Although this interpretation may be open to question, the book nonetheless has an undeniable historical value. The bidding will start at one hundred and eighty thousand pounds."

Trench hesitated for a few moments before reacting, as if doing some mental arithmetic. "I wouldn't be surprised if it sold for more than three times that sum," he said at last, "which would be close to one million dollars." Then he added, with an edge of disappointment: "I'm afraid that's much more than we can afford."

"It doesn't matter. We don't really have to purchase it."

"What do you mean?"

"I mean, all we need to do is find out what's in it; or rather, whether there's something in it that might help our search for Pythagoras."

"And how can we do that if we don't have the book? I don't suppose they give away photocopies of it." Trench was finding it difficult to hide his growing frustration.

"Of course they don't. Potential buyers are allowed to examine the item, but may not take any pictures. They are given only a photocopy of half a page from the book, a brief summary of its supposed contents, and a copy of a lab report certifying the approximate age of the parchment."

"So where do we go from here?" He had calmed down, sensing that Jule might have come up with some way out of the impasse.

"The fellow who wrote the summary is a professor at Oxford," Jule began to explain, and as he went on, Trench's impression seemed to be confirmed. "He must have read and translated the whole book. I wonder whether Laura, as a colleague and author of a book on Pythagoras, may convince him to share with her what he knows."

"I'd be surprised if he did. You know better than I do how things work in academia. The guy will want all the glory for himself, and we'll have to wait months until his article about the manuscript gets published." There was a pause as a thought crossed Trench's mind. He then said: "I've got a better idea. Try to find out all you can about this Oxford professor, where he lives, does he have a family, that sort of thing. What did you say his name was?"

"Elmer Galway."

Trench was smiling as he ended the conversation: "Thanks, call me as soon as you get the information." The thought that had earlier crossed his mind was "This is a job for Houdini."

On a cold and windy February morning, with ominous grey clouds low in the sky above Oxford, Elmer Galway set out to take Slipper for his morning walk. It was their daily ritual, called off only in case of illness or extremely bad weather, and performed with clocklike regularity between 6:15 and 6:45.

For the master, it was an opportunity to reflect on some current problem or project and plan for the day ahead, while the dog, unencumbered with such human preoccupations, was busy barking at other members of his species, chasing a squirrel, lifting his leg, and generally having a good time.

On this particular morning, Galway's thoughts drifted toward the upcoming auction of the Pythagorean manuscript. He wished the Ashmolean Museum could buy it. Then, as the uncontested expert in pre-Socratic Greece at Oxford, he would be given priority to study it and so ensure his paternity over the discovery of a historical bombshell. But even if someone else bought it, he had little to fear. His annotated translation was almost ready for publication, and Green had assured him that no one else would have access to the book before the auction, in a week's time—except for a photocopy of half a page that didn't reveal much of the story. Thanks to his friendship with David Green he had benefited from inside information, and was perhaps guilty of the equivalent of insider trading in financial circles. But that's life, he thought, and this time it was his turn to get lucky.

He was momentarily distracted from his reflections by a strong pull on the leash: Slipper had spotted a squirrel and started on a hopeless chase for the small animal. But after being dragged a few feet, his master's firm grip on the leash steadied the dog, while the squirrel sought refuge on the top branch of a leafless tree. As they turned the corner, Galway paid little notice to the small blue car parked on the other side of the street. Its driver, a wiry young man in his thirties wearing a leather jacket, was intently studying a street map.

The squirrel incident over, Galway resumed his train of thought. He could have submitted his article earlier, but had been held up by his—so far unsuccessful—attempt to make sense of some intriguing drawings and symbols that filled the last eight pages of the medieval book. These pages were no longer part of the book now offered at auction. They had been cut by Señor de Burgos and later stolen from his hotel room; for all practical purposes, they were lost. But the Franciscans who had discovered the book in their basilica had photocopied it, and through Sr. de Burgos he had obtained copies of the original photocopies of the missing pages. Their very poor quality made their interpretation all the more difficult. In those elaborate drawings and geometrical symbols, he had hoped to find some clue that would lead to the location of a papyrus scroll supposedly written by Pythagoras himself. The historical value of such a document, a kind of ancient philosophy's Holy Grail, would be incalculable. He wondered whether whoever stole the parchment pages from de Burgos was also on the trail of Pythagoras' scroll.

As he approached his home at the end of the morning walk, Galway noticed that the low iron gate leading to a small garden in front of the house was open. He always closed it on his way out. As soon as he opened the front door and stepped inside, he sensed something was amiss: a cold draft was coming from the back of the house. All windows were supposed to be shut at this time of the year. The thought flashed through his mind: someone had broken in. He ran toward his study with a barking Slipper leading the way and found one of the two windows wide open and sheets of paper being tossed around by the chilly wind. Otherwise, the room had its normal appearance. When he closed the window he noticed the perfectly circular hole in the glass pane; it had been cut clean to allow the window to be opened from the outside—a professional job, he thought instinctively. He patted Slipper's head to calm him down, and went on to check the rest of the rooms. As far as he could tell, nothing was missing from the living room, dining room, bedrooms, or kitchen. Perhaps the intruder didn't have time to take anything, he thought. It was only when he was back in his study picking up the sheets of paper from the floor that he noticed the empty wooden cabinet under his desk: his computer case was gone.

He then conducted a systematic check of all rooms, beginning with his study. Nothing else was missing. Galway concluded that the thief had specifically come to steal his computer files, or rather some file or files he (or she) was interested in. He had backups for all important files, so the loss of his computer was mostly an inconvenience, except for one thing: his translation of the Pythagorean manuscript was in it. Would that be the information the intruder was after? Regretfully, Galway had to admit that he thought so. It seemed all but certain that he wasn't the only one looking for Pythagoras' scroll.

Chapter 19
WITH A LITTLE HELP FROM
YOUR SISTER

The winter of 1997–98 had been unusually mild and wet around the world, a climate anomaly that meteorologists blamed on the presence of a warm current in the eastern Pacific Ocean known as El Niño. In the Chicago area, the gray, rainy weather had persisted well after spring had set in, but there was not a single cloud in sight the morning of April 15, when Jule and Laura met with Trench and Richter to discuss the state of the Pythagoras search operation. That same morning, one thousand kilometers away, Johanna Davidson was driving from Boston to Montreal to attend a much anticipated mathematics lecture at McGill University.

They were meeting in the Temple's Council Room, where the Order's highest authority held its official sessions. An oval table with eight upholstered chairs around it occupied most of the space. On the wall at the end of the room hung a gold-framed oil portrait of the enigmatic Mr. S, the Order's founder, a silent observer of the Council's deliberations. Jule and Laura stared at the picture of the austere-looking man with the white beard and piercing black eyes before taking a seat on one side of the table. Trench and Richter sat across from them, facing the two large windows through which the morning sun was pouring in. The purpose of the meeting was to take stock of the situation and to plan the future course of action. The mood in the room was gloomy, in sharp contrast with the splendid weather outside.

On one front, progress had been minimal: for all her efforts and numerous trips to libraries and museums, Laura had been unable to identify the place where the drawing depicting the serpent might have come from. She had sought the help of colleagues and other experts—being careful not to reveal the true reasons behind her interest in the information—but to no avail.

The only bright note came from a friend of Laura's at Yale University, Dr. Frieda Schneider, a specialist in Gnostic sects, lending weight to Jule's interpretation of the drawing. After having examined it, she had e-mailed Laura:

"*It's quite possible that your picture decorated the wall of a place of worship. Although unintelligible to the profane, to the members of a sect such as the Naassenes the scene would have evoked their dogma, where the serpent, whom they regarded as the Spirit, grants every living being grace and beauty according to its nature.*

Another reference to the Naassenes is the shape of the lozenge containing the serpent, which resembles that of an almond. A pre-existent almond symbolized for the Naassenes the Father of the Universe."

The lack of concrete results had not dampened Laura's resolve; she still expected her persistence to eventually bear fruit. She told Trench as much: "I never thought the investigation would be short or easy, and I'm certainly not ready to give up yet. Besides Dr. Schneider, who took a particular interest in the affair, and myself, I have two graduate students working on the case. My hunch is that the temple or shrine we are looking for might have been discovered by one of the early German or British expeditions at the beginning of the nineteenth century, complete records of which are very difficult to obtain."

"You have our full confidence, Dr. Hirsch," Trench reassured her. "It never occurred to us to impose any deadline on your work."

"When we hired you and Dr. Davidson, we didn't expect results after a specific period of time," Richter said with an even voice, echoing Trench's words, "but we entertain no doubts as to the ultimate outcome of your team's endeavor: our reunion with the Master reincarnate. It is a certainty we cannot explain; we just know it, that's all."

Developments had hardly been more encouraging on another front. With Houdini's help, and despite their disapproval of the method,

Laura and Jule had been able to get their hands on Galway's translation of the medieval book auctioned in London. According to the Oxford professor, the book was the Arabic version of a letter written by one of the Pythagoreans. Besides relating the circumstances of Pythagoras' death, the letter mentioned the existence of a scroll by the Master himself but gave no indication of its location whatsoever. So, even though this new information about the philosopher's death delighted Laura, who intended to use it in her upcoming book, it was not really helpful as far as the search for his reincarnation was concerned.

If the results were meager, it was not for lack of effort. For the last three months, Laura and Jule had put all their talent and energy into the task. On a typical day, when they were not away doing research or following some lead, they would arrive at Richter's place by taxi at around 8:30 in the morning and work until 6:00 p.m.; they worked five and sometimes six days a week. They had lunch with Richter in the dining room followed by a short break before resuming work for the rest of the afternoon. At the end of the working day, a taxi would take Jule back to his hotel and Laura to her cousin's place. Emma Hirsch was not a real cousin but was descended from a branch of the Hirsch family that had emigrated to the United States in the early 1900s. She worked as a hairdresser and lived alone in a modest but spacious apartment in North Chicago. When Emma offered to put her up, Laura readily accepted the invitation, preferring such an arrangement to staying at a hotel.

On the last day of April 1998, as the taxi arrived at his hotel after their day's work, Jule decided on the spur of the moment to invite Laura for a drink. She hesitated.

He insisted: "C'mon, it's been all work and no pleasure for weeks."

"All right. But you must promise not to order whisky."

"That's easy, I never drink whisky anyway. But why?"

"Never mind, it's a long story."

They got out of the taxi, entered the hotel lobby, and headed for the bar. Shortly after they had ordered their drinks, a bellboy approached Jule: "Mr. Davidson? There's a call for you. You may take it in the booth near the front desk, on your left."

Jule entered the telephone booth and picked up the receiver without bothering to close the door. "Hello."

"I knew I would find you in the bar." Johanna enjoyed teasing her twin brother.

"That's not fair! I haven't had a drink in weeks. I was just unwinding with a colleague."

She wasn't interested in his social life. "I've got news for you—big news," she said. "Did you get my message? I tried to call you on your cell phone but there was no answer."

"I know, the battery is down. What's your big news?"

"It's about a book I'm reading, *Life of a Genius*, the biography of Norton Thorp."

"I know the guy, and I've no doubt he *is* a genius. But what's the connection?"

"I thought you would be interested in reading a couple of passages from the book and then draw your own conclusions."

"Can't you tell me a bit more? I'm not exactly in the mood for biographies at the moment."

"Just trust me. It'll be worth your time. What's your fax number?"

Jule gave it to her. The conversation then moved to other subjects, notably Johanna's recent breakup with Kevin, after a relationship that had lasted a record nine months. Jule listened patiently but didn't say much. He knew better than to interrupt her with well-intentioned but ill-advised "You should have . . ." or "Why didn't you . . ." An attentive ear was all his sister looked for.

Johanna eventually got back to the purpose of her call. "I'll fax you the pages tomorrow morning. Take care," she said before hanging up. The call had lasted some twenty minutes.

Jule hurried back to rejoin his forsaken colleague in the bar. The soft notes of a piano had produced a relaxing effect on Laura, who was leaning against the backrest of her chair with her eyes closed.

"Terribly sorry it took so long," Jule apologized, startling Laura out of her reverie. "It was my sister. She lives in Boston, and we don't see each other very often."

"Everything okay?"

"Oh yes, only she sounded a little mysterious about a book she's reading. She's faxing me some pages she absolutely wants me to read."

"What's the book about?"

"It's a biography; the biography of Norton Thorp, the famous mathematician."

"I've heard about him. He came to Urbana last year to give a talk at the university. It was announced all over the campus, you couldn't have missed it. A high-profile event, with press and TV coverage. He's some kind of modern-day Einstein, I hear."

"I suppose you could say that, except he's interested in mathematics rather than theoretical physics. But his works have changed the face of mathematics just as Einstein's theories revolutionized physics. Actually, some of Thorp's mathematical results have applications to string theory."

"String theory?"

"It's one of the hottest topics in physics. The ultimate constituents of matter, so the theory goes, would be some stuff in the shape of infinitesimally small stringlike filaments, and the elementary particles would be vibration patterns of those strings in spaces of ten dimensions or more."

"You completely lost me," said Laura, almost apologetically. "I must be a romantic, because the word 'strings' only brings to my mind memories of violins and chamber music."

Jule smiled and looked straight into her black eyes. "You are a true romantic, then," he said softly, "but of the quiet type. I hardly know you after all these months of working together."

"Blame it on my former life. When you live under constant surveillance, you quickly learn to tell as little as possible about yourself. You don't trust anybody, not even those you think are your friends. Your feelings, your opinions, even your pastimes, you keep to yourself."

"But you ran away from all that long ago. And besides, you have nothing to fear now; the German Democratic Republic no longer exists."

"Yes, I did escape," she said, staring into space. "But I paid a heavy price, and I made my mother and sister who stayed behind pay a terrible price too."

Her eyes had that characteristic brilliance announcing the imminence of tears.

"I did escape, eventually," she repeated, "but not on my first attempt." There was a silence, as she prepared to tell—and relive—a painful experience. "I was in Vienna, at an international conference. I had a friend at the American Embassy, or rather a friend of my late father, a woman whom I absolutely trusted. She was going to help me make the jump to freedom. Halfway through the conference closing session I was to discreetly leave the conference site, a magnificent eighteenth-century palace in the outskirts of the city. A car sent by my friend would then pick me up and take me to the embassy, where I would become a political refugee. As I was leaving the building, a man approached me and called my name: 'Dr. Hirsch? Your car is waiting, please hurry up.' A car pulled up; we quickly got in and the car drove away at full speed. I realized from the anxious way the man sitting beside me kept looking back that another car was following us. I feared the worst and felt much relieved when we finally managed to shake it off. But just when I thought I was safe, my heart stopped beating: it was not to the U.S. Embassy that I had been taken but to the Soviet Consulate. Someone had betrayed me.

"I'll spare you the details of what happened to me when I got back to Dresden. I spent almost two years in jail. It was an all-female prison, but the warden would occasionally let male officers pay nightly visits to helpless inmates, in exchange for God knows what favors or bribes. One night, I had the honor of being chosen to provide free entertainment for a two-hundred-pound drunken brute. All I remember—and I suppose that's all my brain allows me to remember—is the smell of his breath reeking of cheap whisky as he pressed his sweating flabby face against mine.

"Now you understand why I asked you not to order whisky," she said, unable to control her tears any longer.

Later that night, back in his hotel room and still under the emotional impact of Laura's confidences, Jule had forgotten all about the book his sister had urged him to read. A blinking red light on the telephone panel informed him he had a new message. He took it. It was the message Johanna had left him that afternoon, and it ended with an enigmatic "*I think I've found the solution to your problem.*" "What problem? And why does she want me to read that book?" he thought, remembering his sister's earlier call. He couldn't think clearly, he had had a drink too many. Better wait until tomorrow to sort this out, he decided, and went straight to bed.

When he got to the office the next morning, Johanna's five-page fax was waiting for him. He sat down at his desk and began reading it.

The first page was an excerpt from the introduction of the book. The author claimed to have had access to Thorp's aunt's personal diary, which provided hitherto undisclosed details of the famous mathematician's childhood and youth; among these, two seemingly paranormal incidents involving some strange and inexplicable behavior of the young Norton.

The remaining pages elaborated on the nature of those incidents. The first one took place when Thorp was five years old. One evening, his aunt found him sitting at the piano, playing a piece of classical music with the skill of an accomplished pianist, even though he had never taken piano lessons or even touched a keyboard before. The second bizarre episode occurred some ten years later, when Norton was in high school. The author quoted from an entry in Therese Thorp's diary dated April 28, 1979:

I met with Mrs. Witherington, Norton's literature teacher. She's as bewildered as I am by Norton's homework. The class was studying masterpieces of world literature, and she had chosen a passage from *The Odyssey* relating Odysseus' encounter with the Cyclops, a mythical tale particularly appealing to young readers. The students were to read the chapter and write a short essay on it. She was about to finish reading Norton's essay when she noticed something odd: the last few lines looked like meaningless scribbling. On closer examination, she realized the "scribbling" was an unbroken sequence of Greek letters, six lines deep. She thought she knew what was going on: Norton had, not very adroitly, copied it from some book where the English and Greek versions of the poem were printed side by side. To confirm her suspicion before confronting Norton, she showed the page to her brother-in-law, who taught classical languages at the university. His conclusion, after having discussed the matter with some colleagues, only deepened the mystery. It was Greek all right, and its general sense roughly corresponded to a passage from Book IX of Homer's *Odyssey*. However, it didn't exactly match any of the published versions of the poem he had seen.

She wrote down the English translation for me:

"We came to the land of the Cyclops race, arrogant lawless beings who leave their livelihood to the deathless gods and never use their own hands to sow or plough; yet with no sowing and no ploughing, the crops all grow for them—wheat and barley and grapes that yield wine from ample clusters, swelled by the showers of Zeus."

What is more, her brother-in-law added that the passage was written in an ancient Greek alphabet (Ionic, I think), without separation of words, whereas all known versions of *The Odyssey*—the oldest one dating from the tenth century AD—were written in either Hellenistic or Modern Greek. Therefore, in his opinion, Norton's text had definitely not been copied from any book.

I knew he hadn't copied it. Cheating is certainly not in his nature, and besides he doesn't need to cheat, he always gets the highest mark anyway. I believe what Norton told us, Mrs. Witherington and I. He was working late on his assignment when he fell asleep at his desk. It was well past midnight when he woke up and went to sleep in his bed. The next morning, he picked up the sheets from his desk, convinced he had finished the essay the night before. He turned it in without ever reading it again.

I don't understand, but I gave up trying. I still remember what Morris and I went through ten years ago when we tried to understand. He's an exceptional child, a child like none other; that's all there is to understand.

Jule stopped reading, lifted his head, and whispered "Unbelievable" to himself, in a state of shock and elation at the same time. He remained at his desk for long minutes, sitting still but for a slight motion of his head, after which he turned to his computer and looked at a particular file. Then he got up, went over to Laura's desk across the room and gently tapped on her shoulder to get her attention. When she turned to face him, he very calmly said: "I've found him."

Jule had Laura read the excerpts from Thorp's biography. When she finished, she reached for his arm and pressed it without saying a word—she didn't need to.

"And you know what? There was a clue right under our noses all the time," said Jule.

"What clue?"

"Remember your translation of the Greek poem on page six?"

"Yes. What about it?"

"Are you sure it's accurate, I mean, are the verb tenses right?"

She searched for the file in her computer and a few seconds later the words appeared on the screen:

He is a man of exceptional wisdom
Versed in the secrets of number;
One who of all men
Has the profoundest wealth of intellect.

"Yes, that's correct. As far as I can tell it's a faithful translation of the poem. What's the matter with the verb tenses?"

"Unlike the original verses by what's-his-name . . . "

"Empedocles."

"Right. Unlike his verses, which are in the past tense—I've just checked it—as they should, because he's referring to Pythagoras after his death, this poem is in the *present tense*."

"So?"

"So the poem in the parchment sheet is not really a shorter version of Empedocles,' it is a clue to recognizing the *living* Pythagoras: "He *is a man versed in the secrets of number*, one who *has the profoundest wealth of intellect*; in other words, he is a mathematician of genius!"

Soon afterward, a series of events took place in quick succession. They called Trench—Richter was out of town—requesting an urgent meeting with him. When they presented their case, he became very excited. "Excellent, excellent," he kept repeating.

That evening, the Council of High Companions met to hear Trench's report and make a decision. He lay out before them what he considered to be compelling evidence that Norton Thorp was Pythagoras reincarnate: "a man versed in the secrets of number; one who of all men has the profoundest wealth of intellect" was a fitting description of a mathematician of genius. The little boy who had never touched a keyboard before magnificently playing the piano as if in a trance—the Master was a consummate musician: Wasn't his spirit guiding young Norton's little hands that evening? And finally, the

Greek passage in the Ionian alphabet without separation of words: only an educated 500 BC Greek familiar with Homer's epic poem could have written it.

"I'm aware that any one of these three elements by itself is not sufficient evidence for reaching a conclusion," Trench said to the attentive group sitting around the oval table, "but taken together they constitute as convincing a proof as we can expect to get."

The vote was unanimous in favor of Trench's recommendation, so the second phase of the operation could now be activated: Rocky and Houdini would prepare and execute a plan to bring Thorp-Pythagoras before the Council.

PART IV

PYTHAGORAS' MISSION

Chapter 20
ALL ROADS LEAD TO ROME

In mid-May 1998, Galway received official confirmation that his article on the Pythagorean manuscript had been accepted for publication in the prestigious *Journal of Ancient Philosophy and History*. As for the manuscript itself, it had been bought at the auction by an anonymous collector for the sum of 420,000 pounds. It was not known whether the new owner intended to donate it to a museum or library. That posed a bit of a problem for Galway, who could not include the location of his translation's primary source in his article.

His satisfaction at having his paper accepted was tempered by the fact that he had been unable to crack the secret of the last eight pages of the ancient book, of which he had only very poor quality photocopies of photocopies. He had translated the poem on page six and concluded that it referred to Pythagoras, but that was about all the headway he could claim. He was intrigued by the drawing on page four, showing a serpent and two feminine figures, but couldn't see any connection with the Pythagoreans. The rest of the pages, containing arabesques, mathematical symbols, and geometric figures, seemed to be equally irrelevant as far as the location of Pythagoras' papyrus was concerned. The precious scroll's trail had gone cold.

Disappointed with the lack of results on the papyrus front, Galway turned his attention to his late father's memoirs. Clearly, the old man had entrusted his younger son with the unfinished book knowing he could count on him to complete the writing and have the work published.

Actually, Galway senior had already done most of the work, and had left a printout of the third draft of chapters 1 to 14. Elmer went over it and found that it required only minor corrections and a few additions. In particular, a number of post-it notes reminded the author of facts to be checked, pictures to be inserted, and so forth. Among these, a note in chapter 7 caught Elmer's attention. It read "Neopitagorica Basilica 1935–38—Insert two drawings here." That particular section of the chapter was about the excavation of a very ancient underground basilica situated in the historic center of Rome, and in which Ernest Galway had participated. He had written:

> The building, a large vaulted hall of basilican type, with vestibule, apse, and three aisles divided by pillars, is some fifteen meters long, nine meters wide, and seven meters high. It was discovered in 1917, as a result of a landslide under the roadbed of the Rome-Naples railway, but only in 1935 did serious excavation work begin.
>
> The decoration is elaborate. Walls, vaulted ceilings, and apse are covered with well-preserved stucco reliefs. Among the subjects are mythological compositions, sacrificial and ritual objects, and symbols of resurrection and afterlife. All this suggests that the building was used by followers of some mystic cults which flourished in Imperial Rome.
>
> From the character of the concrete, which contains no fragments of tiles, and from that of the bricks of tufa used in a shaft above a skylight, it may be argued that the building was actually constructed early in the first century AD—and not in the second century, as it is generally believed. Another argument in favor of the earlier date has been put forward by Professor Clermont, who pointed out that the decoration is entirely Greek in spirit, showing no motifs derived from astrology. Clermont also contended that the building was probably used by a Neo-Pythagorean assembly.

His curiosity aroused, Galway looked inside the box labeled "1935–38." It contained photographs, drawings, and miscellaneous records. The drawings had inscriptions on the back (place, date, and so on), so it was easy for him to find those marked "Neopitagorica

Basilica." There were six of them, and one in particular caused his heart to skip a beat: a finely executed drawing of a stucco bas-relief showing a serpent inside an oblong lozenge and flanked by a pair of matching feminine figures—the unmistakable source of the enigmatic illustration in the Pythagorean book. The thought then came upon him with the force of a revelation: the papyrus scroll written in Pythagoras' own hand was hidden somewhere in that basilica.

After doing some research on the Internet, he learned that the Neo-Pythagorean basilica was on the World Monuments Watch list, undergoing a major restoration, and closed to the public. Structural work to the building was also being carried out to fix various problems, such as water permeating the site and an antiquated ventilation system that had favored bacterial growth on the polychromatic surfaces.

He began making preparations for a trip to Rome. Among other things, he would have to cancel a talk he had been invited to give at an international conference in Munich and make arrangements for Slipper to be looked after. But, most important of all, he had to obtain permission to conduct the search from the Italian Ministry of Cultural Heritage. That would likely take weeks, even months, if his experience with bureaucracies around the world was any indication. Fortunately, he knew a junior deputy minister, Dottore Luigi Pisano, through a common friend.

Galway e-mailed Pisano about obtaining the necessary papers and got the following reply:

Caro Professore Galway: Permission to excavate or carry out a search in Rome's archaeological sites is not delivered by my ministry but by the Soprintendenza Archeologica di Roma. Please address your request to them, stating your aims for the excavation or search and for how long (six weeks per year at the most), together with a list of the members of your team and a budget for the work.

It will be helpful if you write directly to the head of the Department of Excavation Projects at the Soprintendenza, Signore Ettore Calabrini, mentioning my name [he gave e-mail and postal addresses]. He will see that your application is put on the fast track.

Welcome to red tape Italian style, thought Galway, and to how to (perhaps) circumvent it by pulling the right levers.

He wasn't sure it was a good idea to tell the Italians the real motive behind his interest in the basilica. He was afraid someone else might get there first, depriving him of the credit he deserved for his role as the discoverer of such a priceless document. But short of staging a commando-type operation to find the papyrus and then smuggle it out of the country, there was no way other than the official way.

He had been mulling over whether to reveal his true intentions on his application for the Italian permit when he got a call from his brother. It was about some papers that needed to be signed in connection with his father's succession. Elmer, who had not mentioned his discovery to anyone yet, could not resist confiding in his brother. He told John about his hunch regarding the location of the papyrus and his apprehension about disclosing too much to the Italians.

Then he said, jokingly: "I even thought of going down in the middle of the night to fetch the scroll without bothering with permits, a mission worthy of that intrepid archaeologist in the movies . . . "

"Indiana Jones. Not a bad idea. It can be done."

"You're not serious, are you?"

"Of course I'm serious. I know some people in Bogotá who specialize in that type of operation, and they have 'branches' all over the world. It won't cost you that much, and results are guaranteed. Absolute discretion too."

Elmer was not really surprised to learn that his brother had contacts in shady circles. In the emerald mining business, protecting a site in the middle of the jungle, when not smuggling gems out of the country, often required using some questionable methods—and the appropriately qualified personnel.

"Are you proposing to steal the scroll?" asked Elmer in disbelief.

"Actually, I was only thinking of borrowing it."

"What do you mean?"

"I mean, suppose you're the one who finds it. If it's so important, you get the glory. Then you give it back to its rightful owner and everybody's happy."

Elmer cut him short, refusing to listen any further. "Thanks for trying to help, but I won't gamble my reputation on some cloak-and-dagger operation."

"No problem. It was only a suggestion." John hadn't really expected Elmer to go along with his scheme. And then he added, changing the subject: "Don't forget to send the signed papers to Harris as soon as possible. He'll forward them to the real estate agent. Good luck with the Italians," and hung up.

Elmer Galway had never smoked and was only a moderate drinker. At age sixty-four, he was in pretty good physical condition, as he had been throughout his adult life. He had a well-proportioned body, with only the hint of a bulge at the waistline, and strong arms, a legacy of his youthful passion for sport. As an active member of the Oxford Rowing Club he still rowed occasionally, but he kept fit chiefly by walking a good deal and strictly following a healthy diet.

This bright picture of his general physical condition undoubtedly played a part in his new approach to the Pythagoras' scroll search-and-recover operation. He still believed the papyrus was hidden somewhere in the Roman basilica and was determined to put his intuition to the test, but he no longer intended to apply for permission to search the building—at least not for the moment. He had a better plan, one that had slowly dawned on him after his telephone conversation with John.

It had been easy for him to obtain detailed information about the basilica (plans, photographs, and so forth) through the World Monuments Watch by manifesting a professional interest in the restoration work being carried out, and he had been glad to see that scaffolds had been put up at various places inside the building.

He had chosen Wednesday, June 17, as D-Day to put his plan in operation. Since he wanted to arrive two days earlier, he had booked a flight to Rome leaving London Heathrow the morning of Monday, June 15.

There was a very good reason for his choice of date: on the evening of June 17, in Montpellier, France, the Italian football team played Cameroon in the World Cup final phase. Not that Galway cared much about football in general or the Italian team in particular, but practically everyone else in Rome did. And therefore, from kickoff time at 9:00 p.m. until the final whistle at around 11:00 p.m., the streets of Rome would be deserted and the city paralyzed, as every single soul in town, from people at home to cooks in restaurants and policemen on duty, would be glued to a TV set watching the game—which suited Galway just fine.

Shortly past noon on the 15th and after an uneventful flight from London, he arrived at Rome's Leonardo Da Vinci Airport. Catching a cab was not a problem: he was solicited by half a dozen taxi attendants competing for attention: "Taxi, taxi to the city, this way please."

He had booked a room in a small two-story hotel in the heart of historic Rome, facing Nero's Aqueduct and a short walk from the basilica. The purported reason for his trip was to look at certain archaeological records at the Vatican Library, but the real purpose of his presence in Rome was to search the Neo-Pythagorean basilica for Pythagoras' scroll.

In the afternoon, after checking in at the hotel and having a light lunch at a nearby trattoria, he headed for Porta Maggiore to perform a reconnaissance of the grounds.

On one side of Porta Maggiore Square, between via Penestrina and Scalo San Lorenzo streets, stands a short wall, its bricks blackened by dirt and grime. It is the wall that supports the railway viaduct carrying the Rome-Naples line. Partly hidden by a recess in the wall and inconspicuous to those passing by, there is a door. Beyond this door, down a flight of stairs and some ten meters below the railway tracks, is an entrance to the so-called Neopitagorica Basilica—although not the original entrance, which is still unexplored.

Galway's main concern was the door lock. He walked nonchalantly along the brick wall looking like the tourist he actually was. When he reached the door, he stopped and examined it for a while with innocent curiosity, and even turned the knob and pushed it. The door was locked, but he was relieved to learn that the lock was a standard one—and fortunately not one of the keyless type. There is really no reason for heavy security, he thought. The building is empty, except perhaps for some scaffolds, and neither graffiti artists in search of notoriety nor vandals in a destructive mood would be interested in showing off their talent or their contempt for society in such a dark and out-of-the-way place.

Having completed the reconnaissance of the site to his satisfaction, Galway spent the rest of the afternoon sightseeing in Rome's historic district. He wondered whether those magnificent vestiges of the Roman Empire's glorious past, such as the Colosseum and the Forum, had yielded all their secrets before beginning a new life as tourist attractions. Perhaps not, he thought, for the site of the Roman Forum was still being excavated and several areas were closed to the public.

He spent the best part of Tuesday at the Vatican Library—he already had a reader's pass from previous visits granting him access to the library's collections. The modern-day library, started by Pope Nicholas V in the 1450s with a few hundred Latin manuscripts, now contained more than 1.5 million printed books and some 150,000 manuscripts in its sixty kilometers of shelf space.

The double-nave main reading room with its beautiful frescoed ceiling was full almost to capacity, but not the slightest sound could be heard. He found an empty chair and sat down at a desk between a monk in a brown robe, his head buried in a pile of books, and a bespectacled young man with a laptop. A fervent lover of books—an attraction that was as much a sensual affair as an intellectual one—Galway wouldn't miss out on the opportunity to hold in his hands some of the library's treasures. He selected a splendidly illuminated survey of Rome's ruins from the fifteenth century by papal secretary and amateur archaeologist Poggio Bracciolini, a prelude to the excavation and restoration of the ancient city that began during the Renaissance.

When he left the library early in the evening, Galway was feeling upbeat and optimistic about the success of his plan. He was also feeling hungry, and for his last dinner before D-Day decided to have a typical three-course Italian meal.

After some searching, he settled for a small restaurant off St. Peter's Square with an outside dining area at the rear. He set aside the menu, happy to follow the recommendations of the waiter, an expansive short man with very shiny black hair who took his order without writing anything down. In response to Galway's request for some typical Italian food, he had earlier explained that there was no such thing as "Italian food" but a constellation of regional cuisines.

Galway's Italian feast began with an entree of *prosciutto* (cured ham) with melon, followed by *cacio e pepe* (cheese and pepper), the simple signature Roman pasta dish, and *involtini di vitello* (savory veal rolls with tomato) as the main course, the whole washed down with a half liter of house white wine. By the time dessert was served, he had had more than enough to eat, but that didn't stop him from finishing the delicious *Baba al rhum*, a traditional Neapolitan pastry soaked with rum and topped with whipped cream.

He dozed off during the taxi ride back to the hotel and had to be woken up by the driver. By midnight he was sound asleep and dreaming he was a nobleman in ancient Rome hosting an all-night banquet.

Unlike the previous two days, when a merciless sun in a whitish blue sky had beaten down on the city, Wednesday morning was overcast and cool. Galway slept late and had breakfast served in his room: orange juice, toast, marmalade, and tea—almost what he had at home, with only the porridge missing.

As the moment to put his plan into execution neared, he grew increasingly restless and decided to kill time by taking a bus tour to the nearby town of Tivoli, famous for its well-preserved Emperor Hadrian's Villa from the second century and its spectacular gardens. After the tour he went back to the hotel and tried to take a nap but couldn't sleep. He forced himself to lie in bed nevertheless. At seven o'clock he got up, opened the shutters, and looked out the second-floor window. The weather was unstable. Heavy gray clouds covered the sky while treetops were swept back and forth by gusts of wind. It looks like it's going to rain, he thought, and that's just fine: the nastier the weather, the fewer the people on the streets.

He wasn't hungry—or was too nervous to feel hungry—so he decided to skip dinner and began packing his material for the operation: coverall, helmet with lamp, archaeological field tools (trowel, brush, etc.), digital camera, and first-aid kit. Everything fit nicely inside the small backpack. He then dressed as he would for a night out to a fancy restaurant, except for the heavy-soled shoes and the woolen sweater he was wearing under his shirt.

At 8:45 he came downstairs. In a small room adjacent to the lobby, a few guests and the front-desk clerk were gathered in front of the TV set. The latter caught sight of Galway as he was sneaking out: "You're not watching the game, Signor Galway?"

"I'm afraid not, I have a dinner appointment," he replied, in as casual a tone as he could manage. And, before stepping outside, he turned and shouted an enthusiastic *"Forza Italia!"*—the rallying cry of the Italian football fans.

The weather remained menacing, but in the end it didn't rain. Galway approached the door walking briskly close to the wall and holding the backpack by the straps in his left hand. There was not a soul in sight on his side of the street. Traffic was minimal. A couple of cars were stopped at the red light on via Penestrina facing him, and a lone and almost empty bus was circling the Porta Maggiore roundabout. As his right hand reached for the door handle, he could feel his heart pounding. He turned the handle and pushed the door. It opened onto a dark corridor, barely illuminated by the glare of the street lamps. Galway quickly stepped inside and closed the door behind him. He remained motionless in the pitch black darkness, leaning against the door while trying to catch his breath.

The first part of his plan had worked to perfection—thanks to John. His brother had made the necessary arrangements for a "specialist" to pick the lock that evening, clearing the way for him to get into the basilica.

A couple of minutes later, he was ready for action. The place was damp and stuffy. He reached for the flashlight in his pocket and switched it on, sweeping the space in front of him with a beam of white light. The narrow corridor ended in a brick wall some twenty feet away, but he could see the opening of a staircase down the corridor on his right.

He changed into his coverall, put on his headlamp, shouldered his backpack, and slowly began going down the stairs, supporting himself with his left hand against the wall. Treading carefully on the slippery surface, he went down until he came to a landing, beyond which the staircase changed direction. After descending another flight of stairs he reached the ground: he was standing on the damp floor of the ancient basilica, his nostrils and lungs filled with the musty air trapped underground in the poorly ventilated building.

He inspected the floor first. It was a mosaic floor, much of which had been preserved, but several rectangular spaces where the pavement was missing suggested that its finest portions, most likely panels with figure compositions, had been removed.

Next, he pointed the flashlight in various directions and toward the ceiling. Moving with caution, he then proceeded to survey the place. The building was rectangular in shape, measuring about fifteen by ten meters, with a vaulted ceiling some seven meters high. Its architecture was of basilican type, the most popular form of early Christian churches: a nave flanked by aisles separated from it by rows of pillars, with a vestibule at one end and an apse at the other. The place was rich in decorations: walls, ceilings, and pillars were all covered with stucco bas-reliefs on a variety of subjects, from mythological scenes to symbols of resurrection and afterlife. In the main chamber all decoration was in white stucco, while the vestibule had a dado of Pompeian red ornamented with figures of flowers and birds, and a ceiling adorned with squares of sapphire blue.

And then he saw it. On the vertical face above the arched opening between two pillars, a serpentlike figure in relief seemed to come to life as its shadow shifted under the sweeping motion of the flashlight beam. There it was, the original of the scene in the Pythagorean book showing a serpent flanked by two feminine figures that had long puzzled him. If his intuition was correct, the papyrus in Pythagoras' own hand could not be far away.

Almost one hour later, an exhausted and disheartened Elmer Galway, his clothes damp and soiled, was beginning to contemplate the possibility that his carefully planned search operation might end in failure.

He had assumed that the papyrus would be hidden in a cavity or hole in the wall somewhere near the serpent bas-relief represented in the ancient book. As was common in Roman buildings of the time, the pillars and arches were built of bricks made of tufa, a volcanic stone with a coarse, porous texture. He had examined the brick wall around and near the bas-relief for some particular sign or peculiarity, anything that might suggest that a treasure was hidden behind, but had noticed nothing special.

When he encountered stucco decorations, he had gently tapped all around the figures with the handle of his trowel, listening for some hollow sound that might indicate an empty space or chamber underneath the stucco layer—but knowing full well that detecting a cavity with such a hopelessly primitive method was a long shot at best.

In order to reach the upper part of the pillars and the wall above the arch, he had climbed on one of the scaffolds that had been put up during the restoration work, presently interrupted for lack of funds. Pushing the heavy structure over the uneven and muddy floor to position it on the right spot all by himself had required considerable physical effort, and after completing the operation he had to stop and rest for long minutes.

The fatigue he was now feeling no doubt contributed to his growing discouragement. He sat down on a half-rotten and wet wooden box and leaned back against one of the pillars to consider his options.

Meanwhile, down at La Mosson Stadium in Montpellier, Cameroon was proving a tough nut to crack for the heavily favored Italian side. True, Luigi Di Biagio had opened the score in the seventh minute and Italy was still up 1-0 well into the second half. But the Cameroonian attack led by Samuel Etoo was tearing Italy's defense apart with increasing ease, forcing the Italian goalkeeper Gianluca Pagliuca to perform save after spectacular save to keep his team ahead.

Back in the underground basilica, Galway was reflecting on the options available to him. Was the papyrus hidden behind the serpent bas-relief itself? There was no way to find out other than piercing through the stucco, which would cause irreparable damage to the 2,000-year-old exquisitely crafted work. But how could he possibly destroy one of the oldest extant bas-reliefs of ancient Rome just to verify a hunch? Extending the search to other parts of the building appeared as an even less attractive alternative, especially in his present condition. He looked at his watch: 10:15.

As he lowered his head almost conceding defeat, the beam from his helmet lamp hit the soiled base of the pillar in front of him at a particular angle. If the angle had been slightly different, he never would have seen what he did. When his eyes, which had been staring into space for a while, focused on the spot illuminated by the light, he jumped to his feet and knelt down on the floor covered in rat droppings next to the pillar. But the image he had just seen had disappeared, and for an instant he thought he had hallucinated. This time using his flashlight he pointed it to a brick at the base of the pillar, slightly moving it to change the angle of the beam of light until the image he had previously seen reappeared in front of his incredulous eyes: ten barely perceptible small dots engraved in the stone and arranged in the shape of a triangle—the Tetraktys, the Pythagoreans' sacred symbol. Two millennia of erosion had all but erased the small, round incisions in the porous stone, but enough of a trace of them remained—enough for the trained eye of an archaeologist to notice them.

No longer feeling his fatigue, he reached into his backpack for chisel and hammer. Working frantically despite the uncomfortable crouching position, he succeeded in first loosening and then removing the stone block, which was slightly larger than a shoebox. The block had been lying on a rectangular stone slab which he had no difficulty in prying up and removing to reveal a flat, sandy surface.

Perspiring heavily and anticipating with excitement the imminent climax, he plunged his trowel into the moist sand. The tool went in only halfway before hitting a hard surface. Buried in the sand was a terracotta jar or flagon, its surface coated with a kind of wax, lying on its side. Its mouth was covered with a piece of cloth tightly tied with a string.

He extracted the jar from its sandy tomb and examined it. It was orange in color and quite heavy, with a broad biconical body, a handle, and a wide cylindrical neck. He didn't bother fetching his measuring tape and estimated its height to be about 40 centimeters. Using a brush, he cleaned the sand-encrusted string and then cut it with his knife at various places until it loosened its grip on the piece of cloth. Very carefully, he removed the cloth lid with the string still attached to it.

The jar appeared to be full of sand, which explained its heavy weight. With trembling hands, he slowly tilted it. Nothing came out; the sand was too tightly packed inside. With the help of his knife he was able to loosen the sand plug. Then, holding the vessel upside down with both hands, he shook it vigorously until sand finally began to pour out.

There was something else besides sand inside the jar: a metal tube—of the kind used to keep papyrus scrolls in ancient times, he thought, convinced by now that it contained what he had come looking for. The tube was in fact a cylinder, apparently made of lead, fitted tightly inside a much shorter one that served as a lid. All around the edge of the lid was a dark substance, possibly bitumen that had been used to seal the container. The two sections were stuck together, but by using his knife and repeatedly twisting and pulling, he managed to remove the lid. Inside the tube was something that looked like a rolled newspaper, only much shorter. He slowly took it out, gently holding it with his tweezers. It was a papyrus scroll, loosely rolled and looking miraculously intact after 2,500 years.

Fifteen meters above the floor of the basilica, Rome erupted into a single and prolonged roar of joy—though not in celebration of Galway's discovery. In Montpellier, Christian Vieri had scored in the seventy-fifth minute, sealing for all practical purposes Cameroon's fate, to the relief of the Italian fans. The Italian striker was to repeat the feat fourteen minutes later and put the icing on the cake in Italy's 3-0 victory over the strong African side.

It was five minutes past eleven o'clock when Elmer Galway stepped into the street and swiftly closed the entrance door to the basilica behind him. In his backpack, he was carrying his soiled coverall and tools. As for the precious papyrus scroll, he had only taken a look at the first column of text, since unrolling it without proper precautions could have seriously damaged it. Satisfied that the papyrus was actually what he had expected, he had taken several photographs of it and its hiding place, after which he had put it back exactly where he had found it and had covered as best he could all traces of his nocturnal visit to the ancient basilica.

The streets, so quiet on his way over from the hotel, were now bustling with activity. A car with an Italian flag floating in the wind and carrying a noisy party drove past him. In neighboring streets, bars and cafés were packed with elated fans and drivers were honking their horns to salute Italy's victory. The loud celebrations will likely continue well into the night, thought Galway. That suited his mood perfectly; he too had something to celebrate.

Chapter 21

KIDNAPPED

The bedside clock read 11:20 p.m. James Parker, better known as Jim or Houdini, was lying in bed on his back, fully clothed, hands clasped under his head, reflecting on the day's events. What should have been a smoothly executed operation had become a complicated one. He could blame it all on Rocky for mishandling Mr. T, but what was the use of laying blame at this point?

Houdini remembered the meeting less than one week ago in Dr. Trench's office. Dr. Trench had then stressed the importance of the operation for the members of the Order, who had long been waiting with anticipation for the moment they would be united with their Master reincarnate. Recently, after a search that had lasted many years, it had been established beyond the shadow of a doubt that their Master had reincarnated as Norton Thorp, a famous mathematician presently living in New York.

But Dr. Thorp was not aware of his true identity, Dr. Trench had explained to them, or of his destiny as a spiritual leader. However, if they, Houdini and Rocky, succeeded in bringing him to the Temple, the evidence that would be presented to him coupled with the heartfelt devotion of his followers would eventually win him over and persuade him to become their Master and guide.

"Dr. Thorp should be treated at all times with the respect due to a sacred person," Dr. Trench had told them. And, looking Rocky straight in the eyes, as if the recommendation were addressed particularly to him, he had added: "Persuasion is the key, with coercion to be used as a last resort and applied only with extreme restraint." Then he had handed them a letter for Thorp, explaining the situation and inviting him to the Temple to meet his future disciples.

Houdini then recalled the events leading to their current predicament.

Very early that morning, they had parked their minivan across the street from the Manhattan apartment building where Thorp lived. When he had come out of the building around noon, they had crossed the street and approached him.

Houdini was now playing back the scene in his head, and how he had described it afterwards to Trench on the phone:

"That was when Rocky made his first mistake, by grabbing Dr. Thorp's arm while I was talking him into coming with us nice and quiet. Of course the guy panicked; he thought it was a mugging attempt and didn't listen to a word I was saying.

"Things only got worse when Rocky applied an arm hold and lifted him off the ground, all this in plain view of the doorman, who by then was calling the police on his cell phone. We had no choice but to get the hell out of there in a hurry. Rocky dragged Dr. Thorp and shoved him inside the car, his big hand pressed against the poor fellow's mouth and nose so tight he almost suffocated to death. Once inside we put tape over his mouth and tied him up. By this time passersby had begun gathering around the vehicle. Luckily they couldn't see what was going on inside thanks to the tinted windows.

"I drove away as fast as I could, merging into the traffic. I was thinking: we've got to get rid of the minivan; the police must've gotten a pretty good description of it from the doorman."

They had then driven across the Hudson River to New Jersey and checked into an out-of-the-way motel off Interstate 80, near Paterson. Rocky had stayed in the van with Thorp while Houdini had asked for a double room with two beds at the back—"to cut down the noise from the traffic"—and had paid in advance for the night.

The one bright spot in their difficult situation was Thorp's change of behavior: after his initial resistance, he had calmed down as if accepting his fate, at least temporarily, and had obediently gotten out of the van and in the room. Whether this was a result of Houdini reassuring him that they meant no harm and that he was actually their guest was hard to tell.

After checking in, they had parked the minivan in front of their room. Rocky had carried Thorp in his arms, but once inside he had untied him to allow him to use the washroom, warning him not to lock the door or remove the tape from his mouth.

When Thorp came back into the room, Houdini had signaled him to sit down and handed him the letter from Trench. After Thorp had finished reading, Houdini had asked, in a hopeful tone: "You understand now? Are you ready to come with us quietly?"

Thorp had reached for a pen and a sheet of paper lying on the night table and had scribbled down his answer: "*I think you're completely crazy. Let me go now and I won't press any charges.*"

Houdini hadn't been annoyed by Thorp's response. "Sorry, man, no deal. We're just following orders. You discuss that with Dr. Trench," he had said with indifference, but the way he had crumpled the piece of paper and buried it in his pocket betrayed a certain frustration.

At 5:15, Houdini had phoned Trench. He had waited until then, hoping for some scrap of good news to report, but he couldn't put off the call any longer.

Trench was obviously not very happy with the situation but had kept his cool. He had listened patiently, making no comments while Houdini spoke. "I'll call you back," he had said, laconically, putting an end to the conversation. "In the meantime, make sure there are no compromising traces left in the minivan."

Thorp's "captors" had removed the tape from his mouth but had tied him to his chair. He hadn't said a word, and kept looking at them with a mixture of amusement and contempt.

They had fried chicken delivered to the motel and had eaten in almost total silence, watching TV. The kidnapping had made the evening news. A description of the minivan—a gray Ford Windstar—and of the two men—6 feet 6, 250 pounds, long brown hair, built like a wrestler; 5 feet 4, 150 pounds, crew cut, skinny—was given. According to one witness, the minivan had Alabama plates, but the number she gave to the police did not correspond to any vehicle registered in that state. Houdini smiled on hearing this—he had done the right thing in manufacturing phony license plates.

The cell phone had rung at 6:45.

"Here's what we'll do," Trench had announced, getting quickly to the point. "Richter is flying to La Guardia first thing in the morning. He'll rent a car at the airport and drive to the motel. There you'll swap vehicles. You, Rocky, and your guest will drive over here in the rented car. Richter will bring the minivan back. If he's intercepted by the police, he won't match the description, and if the vehicle is searched, they won't find anything incriminating inside—you better make sure of that, by the way."

"Sure, boss, we already did. We even vacuumed the interior and threw away the filter," Houdini had said proudly. And then he had added, knowing that it wasn't quite true but eager to reassure Trench: "I don't anticipate any more trouble. Mr. Thorp is behaving now, he really is."

They had watched TV until eleven o'clock and then prepared for the night. Thorp, tied but not gagged, was to sleep in one of the two beds, while they would take turns keeping watch over him. Rocky would take the first shift. Houdini had then slumped into bed not really expecting to sleep, his mind too busy going over the events of the day.

The next morning, they had already checked out and were ready to leave when Richter knocked on the door shortly before noon. He avoided looking Thorp in the eye. A flash of guilt shot through him, but the feeling didn't last. This isn't really a kidnapping, he told him- self; he'll understand, once we have the chance to talk to him at the Temple.

They exchanged only a few words before getting in their respective vehicles. The rented car, a light blue Lincoln Continental with leather seats and tinted windows, left first. Rocky was sitting in the rear with Thorp, who had his hands tied in front of him and his mouth taped shut to prevent any unpleasant incident. The minivan would follow them some thirty minutes later, a gap between the two cars to be maintained throughout their trip back to Chicago. Houdini had switched the phony plates back to the vehicle's original ones.

It was late in the afternoon, as they were driving through Pennsylva- nia, when Thorp saw the highway patrol car parked on the shoulder, some three hundred feet ahead of them on their side of the road. He had been waiting for an opportunity to make a move and quickly decided this was it; there was certainly a risk, but it was worth the chance of attracting the policeman's attention. From where he was sitting, he could not have seen the tractor trailer in the fast lane approaching at full speed and getting ready to overtake them; otherwise, he might have reconsidered his decision.

Thorp had visualized the scene in his head, and now he was ready to play it. In a continuous motion, he slid down along the back of his seat, bent his knees together to a crouching position and then thrust both feet forward with all his strength against the headrest of the seat in front of him. The unexpected blow on the back of his head caused Houdini to momentarily lose control of the vehicle. While Rocky let out a "What the hell . . . ," the startled driver instinctively swerved left and into the path of the oncoming eighteen-wheeler.

Speeding at more than seventy-five miles per hour, the fast-moving truck crashed into the side of the car with the force of a runaway train. The impact ruptured the fuel tank of the Lincoln and sent it spinning into the ditch, where it landed on its roof and immediately burst into flames.

The column of black smoke rising straight up in the still air could be seen from miles away. When Richter caught sight of it against the setting sun, he had a distressing premonition and immediately called Houdini on his cell phone. No answer. He felt a knot in his stomach.

As he approached the scene of the accident, traffic slowed to a crawl. He could see the blinking red lights of police cars and fire trucks in the distance, but it took him an eternity to get close enough to inquire about what had happened. Somebody told him. The words kept echoing in his head: "A blue car, New York plates," "Three people inside," "No survivors."

He drove on past the charred remains of the Lincoln and stopped at the first service station he found. There was a small diner next to it. He went in, ordered a cup of coffee, and headed for the washroom. Throwing up helped him to regain his composure, but he was not ready to break the terrible news to Trench just yet.

Amid his distress at the tragedy, a practical matter popped up in his mind: the authorities would not be able to trace the rented car to him.

The credit card and driver's license he had used were fakes. They had been "cloned" by Houdini from information he had hacked from sites on the Internet. A certain John N. Lewis from Cleveland will have some explaining to do to the police, he thought.

After drinking his coffee he felt better, even hungry. He reached for his cell phone and called Trench.

Chapter 22
THE LAST PIECE OF THE PUZZLE

The journalist from *The Times* had been waiting for half an hour when Elmer Galway rushed into the Senior Common Room clutching a pile of papers and file folders to his chest with one arm. He stopped for a moment to catch his breath, looked around, and then headed toward his visitor.

"I'm terribly sorry for being so late, Mr. Morrison," he said, extending his free hand. "I was held up at a very important meeting; the fate of our scholarly journals was at stake."

Thomas Morrison flashed an understanding smile, got to his feet, and shook Galway's hand. He was a gentle, mild-mannered fellow in his thirties who had joined the London newspaper only recently after struggling many years as a freelance writer. On reading the press release—"*English-Italian team discovers 500 BC Greek mathematical papyrus*"—he had immediately contacted Galway asking for an interview, one of the numerous such requests the Oxford professor had received over the past two weeks.

"It's quite alright, Professor Galway," he said in response to Galway's apology. "I assumed you would be very busy since the announcement of the discovery and didn't expect to get to interview you this soon anyway."

"It's been pretty hectic indeed," Galway admitted, "but I don't mind. I'm happy to see archaeology and history instead of physics and cosmology make news for a change. All those stories about teleportation, dark matter, and multiple universes may well catch the public's imagination, but where's the evidence? They're little more than pure speculation, if you want my opinion."

Morrison hesitated and finally chose not to comment for fear of being dragged into an unwanted confrontation with the professor on the issue. Galway, who hadn't expected a reaction from the journalist, was already leading the way to a quiet corner of the room, where he invited Morrison to sit down before slumping into an armchair.

"Do you mind if I record our conversation?" asked Morrison, producing a small tape recorder from his briefcase.

Galway replied that he didn't and the interview got under way.

"Could you first tell me the circumstances leading to the discovery of the papyrus?"

The professor shifted in his chair and began his story.

"We, that is, my Italian colleague Doctor Antonio Marcheggiano and I, first learned of the existence of the papyrus from a medieval book that surfaced in London last autumn and was later auctioned. I was called in to give an expert opinion on the authenticity and historical value of the book, a thirteenth-century Arabic text on parchment. To the best of my knowledge, the book was a translation of a Greek letter written around 500 BC by one of the disciples of the famous philosopher and mathematician Pythagoras to another of his followers. Among other things of enormous value to historians, the letter mentioned the existence of a manuscript in Pythagoras' own hand, which at the time, 500 BC, could only have been a papyrus scroll. Although the letter gave no indication of its content or possible location, it did mention that the manuscript should be 'protected at all costs.' We therefore concluded that some extraordinary precautions might have been taken to preserve the papyrus from the ravages of time, increasing the probability that it still existed after 2,500 years."

Galway paused and again shifted in his chair.

"And you had no idea of where the papyrus might be hidden?" asked Morrison.

"Not the slightest idea; and the story might well have ended there, had it not been for the discovery of another book, or rather, another part of the same book."

"What do you mean?"

"It turned out that the medieval book originally contained eight more pages of elaborate artistic drawings. These pages had been cut out, presumably to be sold separately, but we managed to obtain photocopies of them. We strongly suspected that those skillfully crafted drawings, arabesques, and mathematical symbols contained cleverly concealed clues to the location of Pythagoras' papyrus. Problem was, despite all our efforts we were unable to decipher the hidden information and began to doubt there was any information at all."

Galway then went on to tell Morrison about his discovery of the sketch of the serpent bas-relief among his late father's papers, which, combined with the drawing they had earlier unsuccessfully tried to decipher, all but revealed the place where the papyrus was hidden. After obtaining the necessary permits from the Italian authorities, he and his colleague had searched the Neo-Pythagorean basilica at the appropriate spot and made the sensational discovery: a 2,500-year-old papyrus, believed to have been written by Pythagoras and miraculously well-preserved.

"And what is the papyrus about?" asked the journalist.

"We're still studying it, but can say for sure that it contains many results of Greek mathematics until now only known indirectly, through writings and translations dating from centuries after Pythagoras' time. It is the oldest extant record of some Greek mathematical discoveries, and as such, a document of incalculable value for the history of mathematics."

"I suppose the vast majority of papyri from Pythagoras' days, if not all of them, were lost or destroyed in the course of time because no particular care was taken to protect them," said Morrison, as a preliminary to his next question: "What's so special about this one to have warranted such extraordinary precautions to preserve it?"

For the first time during the interview, Galway hesitated. He took off his glasses and rubbed his eyes before replying, looking as if he were carefully choosing his words.

"Actually, that's still a bit of a mystery for us. But we're working on it."

Later, back in his office, Galway reflected on how close to (or far from) the truth was the story he had told the journalist from *The*

Times—and others who had asked him essentially the same questions about the discovery of the papyrus. He had only lied by omission, he told himself, for the story was strictly true, except for the fact that his Italian colleague became involved only at the end, and not from the beginning.

On his return to Oxford after his successful Roman adventure, Galway had contacted Antonio Marcheggiano, a well-known Italian archaeologist and director of Rome's Capitoline Museums. He had told the Italian he was on the trail of a papyrus scroll supposedly written by Pythagoras and had lately come across certain records pointing to its possible location, somewhere in the center of Rome. And then he had made his offer: Would Dottore Marcheggiano be interested in becoming his partner in the search for the manuscript in exchange for helping him to obtain the necessary excavation permits from the Italian authorities?

It was the best thing to do, he had thought then and he still did; going at it alone was too risky. Even if he were the one to discover it, the scroll would be immediately seized as belonging to the Italian cultural heritage and he would lose control over it. He doubted the Italians would allow a foreign archaeologist to be the first to study the papyrus, let alone publish a translation of it. To be sure, by getting Marcheggiano on board as co-discoverer he would have to share the glory, but it would guarantee him first-hand access to the precious scroll from the start. Sharing the catch had appeared to him as the only way of securing a piece of it for himself—and a non-negligible one at that.

His plan worked just as he had imagined. The "discovery" of the scroll took place in the presence of Marcheggiano and his team, under powerful lights and in front of video cameras. With appropriate precautions, the papyrus was removed from its metal container, placed in a special protective case and sent to the Capitoline Museums restoration laboratory, where it was carefully unrolled in optimal temperature and humidity conditions, digitally scanned, and subjected to various tests and analyses. A press conference was held, press releases in various languages issued, and a party thrown to celebrate the momentous discovery. Throughout it all, Galway shared the spotlight with the Italian archaeologists: his gamble had paid off.

A three-person team headed by Galway was set up to translate and analyze the ancient text. Fully unrolled, the papyrus measured 24 by 96 centimeters. The scroll was written in columns of about 7 centimeters, continuously, without breaks between words; a space of about 1 centimeter was left between columns. As was customary in ancient times, there was no title at the beginning of the manuscript, and a blank space was left to protect the first part of the roll. It was not signed, and the author did not identify himself either, but it was clear from the style and the quality of the language that whoever wrote it belonged to the elite of highly educated persons. Moreover, the extent and depth of the mathematics suggested that its author was well versed in that science. All in all, and considering also the clues that led to its discovery, the members of the team strongly felt that they had come into possession of an extremely rare jewel, one which until then they had thought did not exist: a manuscript written by Pythagoras himself.

It began with an exhortation:

> All inhabitants of cities or country should in the first place be firmly persuaded of the existence of divinities as a result of their observation of the heavens and the world, and the orderly arrangement of beings contained therein. These are not the productions of chance or men.

This divine "orderly arrangement," the author claimed, was ruled by the pervasive presence of "Number," not only in "the heavens and the world" but also in all human activities and creations. Of the universe, the text said that "*it has existed from all eternity and will remain eternally,*" and that it is made of five "*mathematical*" solids. These solids are at the origin of the four elements (earth, fire, air, and water) of which everything is composed: earth arose from the cube; fire, from the tetrahedron; air, from the octahedron; water, from the icosahedron; and the "*sphere of the All*" from the dodecahedron. Therefore, what the author called "mathematical" solids are the five regular solids, also known as Platonic solids. These are defined as convex polyhedra whose faces are congruent regular polygons, such as triangles, squares, or pentagons.

There followed what amounted to a list of mathematical results furthering the notion of order and harmony and usually attributed to the Pythagorean school: properties of various classes of numbers, "perfect," "triangular," and "square" numbers;* theorems about triangles, polygons, and circles, the famous Pythagorean theorem among them, and a method for constructing right-angled triangles with any given odd number n as the smallest side, which, in modern notation, is expressed by the formula $n^2 + [(n^2 - 1)/2]^2 = [(n^2 + 1)/2]^2$ (for example, when $n = 3$, one obtains the well-known 3, 4, 5 triangle).

The manuscript also mentioned *"the most perfect proportion, consisting of four terms, and properly called musical,"* and illustrated it with the numbers 6, 8, 9, 12. The author was probably referring to the fact that the ratio 6 : 12 between the extremes represents the octave (= 1 : 2); both 6 : 8 and 9 : 12 represent the perfect fourth (= 3 : 4), and 6 : 9 and 8 : 12, the perfect fifth (= 2 : 3). Moreover, 9 is the arithmetic mean of the two extremes ($9 = (6 + 12)/2$) and 8 their harmonic mean. (Given two numbers a, b, their *harmonic mean* is the number c equal to $2ab/(a + b)$—or, equivalently, $1/c$ is the arithmetic mean of their reciprocals $1/a$ and $1/b$).

The scroll ended with an enigmatic warning against *"the coming of a false prophet seeking to deceive and mislead men by preaching the preponderance of chaos over order and harmony, and the rejection of Number as the unifying principle through which a perfect knowledge of all that exists can be attained. He must be stopped, and with the help of the Gods, he will."*

The meaning of this last part remained obscure to the team that studied the papyrus. If one were to ignore it, they argued, the ancient manuscript could be considered a kind of compendium of Pythagorean principles and discoveries, which the famous philosopher had taken great pains to preserve for the benefit of future generations.

Galway reluctantly rallied to this interpretation, which was the one favored by the two other members of the team, but he had the nagging feeling that something wasn't entirely right, that a piece of the puzzle was missing.

On a dreary, chilly morning in late November 1998, after Galway had come back from his morning walk with Slipper and was about to get

---

*See appendix 5.

in the shower, the doorbell rang. He wasn't expecting any visitors, especially at such an early hour.

With Slipper barking at his side, Galway looked through the peephole and saw a short, rather young man with a pleasant looking face wearing a blue parka and gray trousers. The notion that the stranger was calling at the wrong place crossed his mind but it was quickly dispelled. No sooner had he half opened the door than the man said, with an eager voice:

"Professor Galway? Please forgive me for imposing myself on you like this but I'm sure you'll be interested in what I have to tell you about Pythagoras."

Something in the man's appearance—the intelligent look in his clear green eyes, perhaps—was reassuring enough for Galway to open the door completely and invite him in.

"Please come in, Mister . . . ?

"Davidson. Jule Davidson." The unexpected visitor stepped inside and immediately became the object of a thorough sniffing inspection on the part of Slipper. Galway, clad in a red-and-white striped bathrobe, led the way down the hall and into the living room. He said, without turning his head: "Have a seat, Mr. Davidson, please. I'll be back with you in a few minutes," and disappeared, leaving the dog behind as if to keep watch over the stranger.

Jule sat down in a battered leather sofa facing the fireplace and studied the room. It was dark, the only light coming through a window that opened on the front garden, since Galway had not turned on the lights. The walls were covered with framed photographs, mostly groups of people; there were a couple of paintings and a series of brightly colored African masks. Furniture was sparse—an armchair and two small tables beside the sofa—but there were plenty of objects crowding the room, from small ivory statuettes to a large Egyptian sarcophagus standing in one corner. Artificial lighting was provided by a table lamp, a floor lamp, and a crystal chandelier.

"Where are you from, Mr. Davidson?" The question startled Jule, who instinctively got up, only to be motioned to remain seated by Galway, who had changed into a tweed jacket and corduroys.

"I'm from the eastern United States; New Hampshire, to be exact."

"And you came all this way to tell me something about our common friend Pythagoras," said Galway in a faintly amused tone. He was sitting in an armchair across the room and seemed to be enjoying the company of his unannounced American visitor.

"Actually, I'm traveling through Europe on a kind of sabbatical. I have plenty of time. By the way, I owe you an apology." Jule leaned forward and fixed Galway: "A few months ago someone broke into your house and stole your computer. I feel partially responsible for that and I'll explain why. Please accept my sincere regrets for the intrusion and the theft."

Galway hadn't expected the conversation to take such a turn. He became a little tense and it showed in his voice.

"You're not the person who broke in, I gather."

"No, that person is now dead."

An awkward silence followed. It was broken by Jule: "Let me start from the beginning. I don't suppose you know an American," he hesitated, "pseudo-religious organization called 'The Beacon.'"

"No, I don't. What about it?"

"They are in fact a sect of Neo-Pythagoreans who worship Pythagoras as their master."

"Esoteric sects of that kind existed during the Roman Empire, but I wasn't aware of any contemporary Neo-Pythagorean group. Are you a member of that sect?"

"No, but I worked for them. They hired me to help them find Pythagoras."

Galway almost jumped from his seat.

"I beg your pardon."

"Yes, that's precisely what they wanted."

"Are we talking about the same Pythagoras?"

"We are. But let me explain. The leaders of the sect claimed to have found reliable evidence in some obscure Middle Eastern libraries that Pythagoras would be reincarnated in the middle of the twentieth century." Jule paused and observed Galway's reaction. The professor remained expressionless, waiting for him to go on.

"I think it was more of an act of faith on their part," Jule resumed. "Whatever the case, they hoped to find Pythagoras reincarnate and convince him to become their master and guide."

"Do you believe in reincarnation?"

"I didn't at the beginning, but . . . some things happened; I don't know what I believe anymore. I was part of the search team. When we learned that a medieval book with references to Pythagoras was for sale in London, we wanted to know what was in it. Since we couldn't afford to buy it, we—I mean, a member of our team—stole your computer, hoping to find the translation among your files."

"And so you did. Congratulations," interrupted Galway, with an edge of sarcasm.

"Yes, but it didn't help our search for Pythagoras."

"And in the end, did you find him?" Galway asked point-blank. He was of course expecting an unqualified "No," but Jule's answer was far from categorical and certainly not in the negative.

"For a while, I thought we had found him, but I'm no longer so sure."

Galway couldn't hide his puzzlement. He moved forward to the edge of his seat.

"What do you mean?" he asked, fixing Jule intently.

Jule then told him about what they had thought was overwhelming evidence that Norton Thorp was the reincarnation of Pythagoras. He didn't mention the tragic kidnapping episode and simply said that, sadly, Thorp had died in a car accident before they could talk to him. In any case, the police had never traced the kidnapping to Trench or the Neo-Pythagorean sect. The death of such a high-profile scientist had prompted a huge federal investigation, but the FBI had concluded that Rocky, already known to the police, had kidnapped Thorp for ransom with the help of an accomplice.

Galway did not say anything for a while. He seemed to be struggling between the impulse to dismiss outright any possibility of reincarnation and acknowledging that perhaps there was something more than simple coincidences in Jule's story.

"You said that you're no longer sure that Thorp was Pythagoras reincarnate. Why?" he asked at long last.

"I recently came across a paper of Thorp's that was published posthumously, a sort of mathematical testament. By the way, you never asked me what I do for a living. I'm a mathematician, so I can fully appreciate the significance of Thorp's landmark result: randomness and chaos, and not predictability and order, are at the heart of mathematics. And this has far-reaching consequences for other sciences as well, physics in particular, for if you can't make connections, you cannot solve or prove things. Such a state of affairs is the exact opposite of the ideas advocated by Pythagoras, for whom it is numbers, and not chance, that rule the world, and by unlocking their secrets one can understand anything. According to Thorp, however, the secrets of numbers are for the most part impenetrable, so Pythagoras' program is doomed from the start. To put it bluntly: if we compare Pythagoras to Christ, then Thorp is the Antichrist, so how could he be Pythagoras reincarnate?"

He paused and noticed Galway's reaction to his words. The professor was overcome with excitement, his mouth open but producing no sound, until he finally said, talking to himself: "The missing piece of the puzzle. . . . *'Beware of the coming of false prophets seeking to deceive and mislead by preaching the preponderance of chaos over order.'*" He was quoting from the last part of Pythagoras' papyrus, which now appeared to him in a new light.

"The prediction of Pythagoras' reincarnation may have been right after all," he said to Jule, who had no idea what was going through the professor's mind.

Galway went on: "Let me tell you how the different pieces of the puzzle *may*, just may, fit together—reconstructing the past is not an exact science.

"Pythagoras somehow learns—an Oracle tells him, or he has a dream, I don't know—that sometime in the future a man of unsurpassed intelligence, admired and respected throughout the world, will claim to have a proof that it is chance and chaos, and not the predictability and order of numbers, that actually rule the world. To Pythagoras and his school, such a person would personify the Anti-Pythagoras—or the Antichrist, as you called him—one who disseminates false and evil ideas to prevent men from comprehending the true nature of reality. This future heretic must be stopped at all costs. Such will be the mission of Pythagoras reincarnate.

"Pythagoras then writes down his most treasured results and the purpose of his mission *as a message to himself reincarnate* and sets up a mechanism for the preservation of this information, maybe through a chain of custodians, each passing on the document to the next. This may explain the extraordinary precautions taken to protect it. Some ancient historians, such as the third-century philosopher Porphyry, mention certain writings that later Pythagoreans left to their sons or wives to preserve within the family, a mandate that was obeyed for a long time.

"At some point, the precious scroll is hidden in the Neo-Pythagorean basilica, perhaps by a member of some esoteric sect who worshipped there, as the best way to protect it, and clues to its recovery and to the recognition of Pythagoras reincarnate are left behind—the letter of the Pythagorean, mentioning a manuscript in the Master's hand; the drawing of the serpent bas-relief; the Tetraktys engraved in the stone; the Greek poem, and maybe others as well. Copies and translations of these documents—not always faithful or complete—circulate for centuries. Most of them are lost or destroyed, but one finds its way into the bowels of the Basilica of St. Francis, in Assisi. When the September 1997 earthquake exposes the medieval book, it sets in motion the chain of events leading to the discovery of the precious papyrus."

Jule had been listening to the professor without daring to interrupt him. He had many questions, but there was one in particular he was anxious to ask: "So you believe that Thorp was probably the Antichrist, or rather the Anti-Pythagoras that Pythagoras so feared, right?"

"I didn't say that. My story is at best a plausible guess and at worst wild speculation. Personally, I don't believe in reincarnation, but if I did, I wouldn't rule out the possibility that Thorp could have been both the Anti-Pythagoras *and* Pythagoras reincarnate, who, thanks to his exceptional intellect and equipped with the knowledge of twentieth-century science and mathematics, demolished the very foundations of his own doctrine."

"If that's the case. . . . What an irony! Being reincarnated as your worst enemy!" Jule shook his head and sat back in the soft, deep leather sofa. He felt tired.

It was already mid-morning, and neither of them had had anything to eat.

"I'm hungry," said Galway getting to his feet. "What about you?" And before Jule had a chance to answer, he added: "Come to the kitchen, I'll cook us some breakfast."

EPILOGUE

"*Cabin crew, take your seats.*" The captain's voice filled the dimly lit business class section of the cabin where Elmer Galway and Bradley Johnston were comfortably seated. British Airways flight 93 from London Heathrow to Toronto's International Airport was on its final approach to its destination. Galway looked out the window. The aircraft was plunging into a sea of thick clouds and a light drizzle tapped on the window pane. It was the spring of 2000, almost two and a half years from the day Irena Montryan had first come to Oxford to seek Galway's help for her exhibition and had ended up talking to his former student instead.

Bradley Johnston was looking forward to seeing her again and hoping to have the chance to get to know her better. The next few hours promised to be busy, though. A limousine was to pick them up at the airport and take them to their hotel. They would then have just enough time to check in, freshen up, and get dressed for the evening before being chauffeured down to the museum for the exhibition's official opening at 7:30 p.m. Many local personalities and officials were expected to attend the reception. Galway would deliver one of the speeches, a foretaste of his talk on the papyrus and the story of its discovery scheduled for the following afternoon.

Bradley, with no official business of his own, would probably be following Galway around and be taken for his assistant—a prospect he didn't particularly relish. Irena had insisted he should come along too, and since the museum covered the expenses he had accepted the invitation. Perhaps she had something in mind for him. The thought of this possibility cheered him up somewhat.

The colorful banner hanging across the Royal Ontario Science Museum's neoclassical façade was hard to miss:

MATHEMATICS: FROM ITS ORIGINS TO ITS FRONTIERS

Inside the building, the exhibition's star attraction could be seen in a separate room, under discreet but nonetheless heavy security. It was displayed, fully unrolled, inside a glass case sitting in the center of the room. Two cones of subdued lighting illuminated the case. The label read:

Earliest extant record of Greek mathematical discoveries.
Written in classical Greek (Ionian alphabet) without separation of words.
Attributed to Pythagoras.
Papyrus, c. 500 BC.
Loan from Capitoline Museums, Rome.

Appendix 1
Jule's Solution

Below, we give some hints for the solution of the probability question posted on canyousolveit.com (p. 6). The concepts, formulas, and notation used should be familiar to an undergraduate student in mathematics.

We consider the general case of n players. A distribution of n hats among these n players is represented mathematically as a permutation s on a set $\{1, 2, \ldots, n\}$ of n elements, where $s(i) = j$ means that player i picks hat j. The number of these permutations is $n!$

Those permutations s for which $s(i) \neq i$ for $i = 1, 2, \ldots, n$, are called *derangements*, and they correspond to distributions of hats for which none of the players picks his own hat.

Then, the probability that none of the n players will pick his own hat is $d/n!$ where d is the number of derangements.

To find the value of d, let P denote the set of all permutations and D the set of derangements (on n elements).

P can be expressed as the union of D and subsets F_i, $i = 1, 2, \ldots, n$, of P where F_i is the set of permutations s that leave i fixed, that is, such that $s(i) = i$. In symbols,

$$P = D \cup \bigcup_{i=1}^{n} F_i.$$

From the equation above it follows that $n! = d + f$, where f is the number of permutations in the union of the F_i.

The collection $F_1, F_2, \ldots, F_n$ of subsets of P has the following property:

The number of elements in the intersection of any k subsets in the collection depends only on k and it is given by $(n-k)!$

Applying the so-called Inclusion-Exclusion Principle to the F_i we get the formula

$$f = \sum_{k=1}^{n} (-1)^{k-1} \binom{n}{k} (n-k)!.$$

Finally, it is straightforward to obtain

$$d/n! = 1/2! - 1/3! + 1/4! - 1/5! + \ldots \pm 1/n!$$

which gives the desired probability.

For $n = 12$, this probability is approximately 0.3679.

The limit of $d/n!$ as n tends to infinity is the sum of the infinite series representation of e^x for $x = -1$, or $1/e$.

Appendix 2
Infinitely Many Primes

Galway had come across Euclid's proof that there are infinitely many prime numbers. Here's a short primer on primes followed by the proof.

Some positive integers can be decomposed into a product of two smaller ones; 28 (= 4 × 7) and 315 (= 9 × 35) are examples of such composite numbers. Those that can't be so decomposed are called *prime*. Put another way: a positive integer $p > 1$ is prime if its only (positive) divisors are itself and 1, so that it can only be decomposed in the trivial way $p = 1 \times p$. Prime numbers are the "blocks" from which all numbers can be built, in the sense that every positive integer greater than 1 is a product of primes (or it is itself a prime)—and it is therefore divisible by some prime. For example, 4,095 = 3 × 3 × 5 × 7 × 13, so 4,095 is divisible by the primes 3, 5, 7, and 13.

The first ten prime numbers are 2, 3, 5, 7, 11, 13, 17, 19, 23, and 29. Does the list of primes eventually stop or does it go on forever? In other words, is the set of all prime numbers finite or infinite? Euclid answered this question in the *Elements*. He stated the theorem that there are infinitely many primes without using the term "infinity": "Prime numbers are more than any given multitude of prime numbers" (Book IX, Proposition 20). His proof is short and beautifully simple. Here it is, essentially unchanged, in modern notation:

Consider the list of primes up to a certain prime P. If we multiply together all the numbers on the list and add 1 to the result, we obtain a number $N = (2 \times 3 \times 5 \times \ldots \times P) + 1$ greater than all those on the list. Now, either N is prime or it isn't. If N is prime, then the list of primes doesn't stop at P. On the other hand, if N is not prime, then it's divisible by some prime, say H. But H must be different from all the primes on the list because none of these divides N (the remainder of such a division is always 1). Thus, in either case we have found a prime not on the list, so the sequence of primes never stops; in other words, there are infinitely many primes.

In September 2006, using a computer program, it was discovered that the number $2^{32,582,657} - 1$ is prime, and at the time it was the largest known prime. This is an extremely large number; if written down, its nearly 10 million digits would stretch for almost 20 kilometers. Some 2,300 years ago and without the help of any computer, Euclid knew that such large prime numbers—and even larger ones—existed: he had *proved* it. Such is the power of mathematical proofs.

Appendix 3
Random Sequences

In 1919, the Austrian-born mathematician Richard von Mises proposed the following definition: an infinite sequence s, say, of $0s$ and $1s$ is *random* if

(a) s satisfies the *law of large numbers*, that is, 'there are as many $0s$ as there are $1s$,' or, more precisely, the limiting value of x/n, where x is the number of $0s$ among the first n terms of the sequence, is 0.5, and

(b) Every subsequence that can be extracted from s by reasonable means also satisfies the law of large numbers.

Applying von Mises definition to the sequence 0 1 0 1 0 1 0 1 ... of alternating $0s$ and $1s$ would confirm that it is not random, for the subsequence of even bits (the 2nd, 4th, etc.), that is, 1 1 1 1 1 1 ..., clearly does not satisfy condition (a). Likewise, many other sequences which appear intuitively to be nonrandom fail to satisfy von Mises' conditions, and hence they are not random also in the technical sense.

Unfortunately, the definition proposed by the Austrian mathematician suffered from a fundamental defect: it did not specify which means for extracting a subsequence are "reasonable" means. To remedy this situation Alonzo Church, an American mathematician, suggested in 1940 that condition (b) of von Mises' definition should only apply to *computable* subsequences, that is, to those subsequences whose terms could be defined by a computer program. Although Church's idea had the merit of making the definition precise, examples were subsequently found of sequences that are intuitively nonrandom but nevertheless satisfy the von Mises-Church notion of "randomness." The modified definition was therefore too large and the concept of a binary random sequence still could not be pinned down.

A Simple Visual Proof of the Pythagorean Theorem

The area of a square of side $a + b$ is $(a + b)^2$ or $a^2 + b^2 + 2ab$ (I).

On the other hand, if the interior of the square is divided up into a square of side c and four right-angled triangles of sides a, b as shown below, its area equals the sum of the areas of these five figures, that is $c^2 + 4(ab/2)$ or $c^2 + 2ab$ (II).

Since the expressions (I) and (II) represent the same area, we have

$a^2 + b^2 + 2ab = c^2 + 2ab$, which implies
$a^2 + b^2 = c^2$

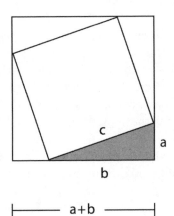

Perfect and Figured Numbers

Perfect Numbers

One classification of numbers attributed to the Pythagoreans involves comparing a number with the sum of its proper divisors—that is, excluding the number itself. If this sum is smaller than the number, the latter is called *defective*; if it is greater, the number is *excessive*. For example, 15 is defective, since $1 + 3 + 5 < 15$, and 24 is excessive ($24 < 1 + 2 + 3 + 4 + 6 + 8 + 12$). A number is *perfect* if it's neither defective nor excessive, in other words, if it is equal to the sum of its proper divisors. Examples of perfect numbers are 6 ($= 1 + 2 + 3$) and 28 ($= 1 + 2 + 4 + 7 + 14$).

In the final proposition of Book IX of his *Elements*, Euclid provided a recipe for constructing perfect numbers: if $2^n - 1$ is prime, then the product $2^{n-1}(2^n - 1)$ is a perfect number. The two examples of perfect numbers above arise from Euclid's formula by taking $n = 2$ and $n = 3$. The next two perfect numbers are 496 (for $n = 5$) and 8128 ($n = 7$).

If you were left with the impression that Euclid's formula will readily lead to the discovery of plenty of perfect numbers, consider this: all the perfect numbers the Greeks ever knew were the four mentioned above, and it was not until the fifteenth century that another one was added to the list ($2^{12}(2^{13}-1)$). Today, only about forty perfect numbers are known, and their fascinating theory still contains many simple but unanswered questions, such as: Is there an odd perfect number? Are there infinitely many perfect numbers?

Triangular Numbers

It was common practice among the Pythagoreans to represent numbers using dots or pebbles arranged in geometrical patterns. The simplest of these are the *triangular* numbers.

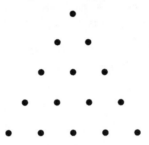

The triangular number 15

Pythagoras is credited with discovering that triangular numbers arise by adding n consecutive natural numbers (starting from 1): $1 + 2 + 3 + \ldots + n$. For example, $1 + 2 + 3 + 4 + 5$ produces the triangular number 15, represented as a 15-dot triangle of side 5, as in the figure above.

Square Numbers

These are obtained when n successive odd natural numbers are added: $1 + 3 + 5 + 7 + \ldots + (2n - 1)$.

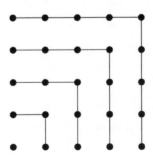

The square number 25

That the above sum always produces a square number can be seen either geometrically (as shown in the figure, where each successive odd number, represented in the shape of an "L," increases the side of the square by one) or algebraically, using the identity

$$1 + 3 + 5 + \ldots + (2n - 1) = n^2.$$

Notes, Credits, and Bibliographical Sources

Main source on Pythagoras and the Pythagoreans:
The Pythagorean Sourcebook and Library, compiled and translated by Kenneth
Sylvan Guthrie, introduced and edited by David R. Fideler. Grand
Rapids, MI: Phanes Press, 1987.

Additional sources:
W.K.C. Guthrie, *A History of Greek Philosophy, Vol. 1, The Earlier Presocratics and
the Pythagoreans*. Cambridge: Cambridge University Press, 1962.
Charles H. Kahn, *Pythagoras and the Pythagoreans: A Brief History*. Indianapolis,
IN: Hackett Publishing, 2001.

Chapter 1

Source on the history of the Fifteen Puzzle:
Y. I. Perelman, *Fun with Math and Physics*. Moscow: Mir Publishers, 1988.

Chapter 2

Illustration (Greek coin): Courtesy of the Philosophical Research Society.

Chapter 4

Proclus' quote from Ivor Thomas, *Greek Mathematical Works, Vol. I, Thales to
Euclid*. Great Britain: Fletcher & Son, 1939, p. 155.
Giovanni Belzoni's book was published in London by John Murray.

Chapter 8

Epigraph from George Johnston Allman, *Greek Geometry from Thales to Euclid*. Dublin: Hodges, Figgis & Co., 1889.

Proof of the incommensurability of the diagonal based on an argument in Lucio Russo, *La Rivoluzione Dimenticata*. Milan: Feltrinelli, 1996, p. 53.

Chapter 10

Flaws in random number generators based on:

Alan M. Ferrenberg, David. P. Landau, and Y. Joanna Wong, "Monte Carlo Simulations: Hidden Errors from 'Good' Random Number Generators," *Physical Review Letters* 69, no. 23, December 7, 1992, pp. 3382–3384.

William Bown, "Gambling on the Wrong Numbers from Monte Carlo." *New Scientist* 24, April 1993, p. 16.

Chapter 11

Thorp's groundbreaking result and its consequences based on:

Gregory Chaitin, "A Random Walk in Arithmetic." *New Scientist*, March 24, 1990.

C. S. Calude and G. J. Chaitin, "Randomness Everywhere." *Nature* 400, July 22, 1999, pp. 319–320.

Marcus Chown, "The Omega Man." *New Scientist*, March 10, 2001, pp. 29–31.

For more on randomness in mathematics, see Chaitin's latest books:

Meta Math!: The Quest for Omega. New York: Pantheon, 2005.

Thinking about Gödel and Turing: Essays on Complexity, 1970–2007. Singapore: World Scientific, 2007.

Chapter 13

Epigraph from K. S. Guthrie, *The Pythagorean Sourcebook and Library*, 1987, p. 117.

Chapter 14

The excerpts from "Song of the Hindu" on pages 89–90 are from *Return of the Aryans,* by Bhagwan S. Gidwani, and they are reproduced courtesy of the publishers (Penguin Books India) and the author.

Chapter 16

Excerpts from the book on the beginnings of Greek mathematics are quotations (slightly edited) from Árpád Szabó, *The Beginnings of Greek Mathematics*.

Dordrecht, Holland / Boston, MA: D. Reidel Publishing, 1978, pp. 186–191.

Chapter 17

The figure showing the illustration from the medieval book reproduced from Jerôme Carcopino, *De Pythagore aux apôtres : étude sur la conversion du monde romain*. Paris: Flammarion, 1956, p. 116, with permission from the publisher.

Chapter 19

Passage from Homer quoted from *Homer: The Odyssey*, World's Classics, Walter Shewring, translator. Oxford and New York: Oxford University Press, 1980, p. 101.

Chapter 20

Description of the Neo-Pythagorean basilica based on:
George H. Chase, "Archaeology in 1917." *The Classical Journal* 14, no. 4 (January 1919), pp. 250–251.

mark 09/09